SpringerBriefs in Applied Sciences and Technology

SpringerBriefs present concise summaries of cutting-edge research and practical applications across a wide spectrum of fields. Featuring compact volumes of 50 to 125 pages, the series covers a range of content from professional to academic.

Typical publications can be:

- A timely report of state-of-the art methods
- An introduction to or a manual for the application of mathematical or computer techniques
- A bridge between new research results, as published in journal articles
- A snapshot of a hot or emerging topic
- An in-depth case study
- A presentation of core concepts that students must understand in order to make independent contributions

SpringerBriefs are characterized by fast, global electronic dissemination, standard publishing contracts, standardized manuscript preparation and formatting guidelines, and expedited production schedules.

On the one hand, **SpringerBriefs in Applied Sciences and Technology** are devoted to the publication of fundamentals and applications within the different classical engineering disciplines as well as in interdisciplinary fields that recently emerged between these areas. On the other hand, as the boundary separating funda-mental research and applied technology is more and more dissolving, this series is particularly open to trans-disciplinary topics between fundamental science and engineering.

Indexed by EI-Compendex, SCOPUS and Springerlink.

Saurabh Sambhav · Deepak Kumar Singh ·
Ashok Kumar Pandey · Azman Ismail ·
Fatin Nur Zulkipli · Andreas Öchsner
Editors

Empowering Solutions for Sustainable Future in Science and Technology

 Springer

Editors
Saurabh Sambhav
Amity University Patna
Patna, Bihar, India

Ashok Kumar Pandey
Employees Selection Board Bhopal
Directorate of Technical Education
Rajiv Gandhi Proudyogiki
Vishwavidyalaya (RGPV)
Bhopal, Madhya Pradesh, India

Fatin Nur Zulkipli
Information Science Studies
College of Computing, Informatics
and Mathematics
Universiti Teknologi MARA
Machang, Kelantan, Malaysia

Deepak Kumar Singh
Amity University Patna
Patna, Bihar, India

Azman Ismail
Maritime Engineering Technology
and Centre for Women Advancement
and Leadership
Universiti Kuala Lumpur Malaysian
Institute of Marine Engineering Technology
Lumut, Perak, Malaysia

Andreas Öchsner
Faculty of Mechanical Engineering
Esslingen University of Applied Sciences
Esslingen, Baden-Württemberg, Germany

ISSN 2191-530X   ISSN 2191-5318   (electronic)
SpringerBriefs in Applied Sciences and Technology
ISBN 978-3-031-77836-0   ISBN 978-3-031-77837-7   (eBook)
https://doi.org/10.1007/978-3-031-77837-7

This Springer imprint is published by the registered company Springer Nature Switzerland AG
The registered company address is: Gewerbestrasse 11, 6330 Cham, Switzerland

If disposing of this product, please recycle the paper.

Preface

In the book *Empowering Solutions for Sustainable Future in Science, Engineering and Technology*, readers embark on a journey into the forefront of innovation where science, engineering and technology converge to address the pressing challenges of our time. Through a captivating exploration of cutting-edge research, real-world case studies and visionary insights, this book illuminates the dynamic forces shaping our collective journey towards a more sustainable and prosperous future. The book also explores how breakthroughs in science, engineering and technology are reshaping our world for the better. This book is beneficial for research scholar, undergraduate and postgraduate students, and industry personnel. The book *Empowering Solutions for Sustainable Future in Science, Engineering and Technology* inspires and empowers readers to become catalysts for a positive transformation in our communities and beyond.

Patna, India	Saurabh Sambhav
Patna, India	Deepak Kumar Singh
Bhopal, India	Ashok Kumar Pandey
Lumut, Malaysia	Azman Ismail
Machang, Malaysia	Fatin Nur Zulkipli
Esslingen, Germany	Andreas Öchsner

Contents

A Novel Health Monitoring System Utilizing IoT and Machine Learning Techniques for Elderly Patient Care

Amit Kumar Mishra, Alok Kumar Yadav, Jagendra Singh, Prabhishek Singh, Manoj Diwakar, and Mohit Tiwari

Abstract Patient monitoring and care delivery can undergo transformative changes through the adoption of Internet of Things (IoT) technologies in the healthcare industry. This study delves into the utilization of IoT for continuous health monitoring of senior patients, a particularly relevant endeavor during situations like pandemics and remote home treatments in intensive care units (ICUs). The research establishes a comprehensive IoT-based healthcare framework by integrating various sensors, machine learning algorithms, and cloud computing. The study encompasses a diverse group of 50 participants, including older individuals in good health and those with various health issues. Artificial neural network (ANN), long short-term memory (LSTM), and decision tree machine learning models are employed to analyze sensor data generated by temperature and pulse rate measurements. These models aim to proactively identify health irregularities, enabling timely intervention. The findings

A. K. Mishra
Department of Computer Science & Engineering, Graphic Era Hill University Dehradun, Dehradun, India
e-mail: amitmishraddun@gmail.com

A. K. Yadav
Naraina College of Engineering and Technology, Kanpur, India
e-mail: its4alok@gmail.com

J. Singh (✉) · P. Singh
School of Computer Science Engineering & Technology, Bennett University, Greater Noida, India
e-mail: jagendrasngh@gmail.com

P. Singh
e-mail: prabhisheksingh88@gmail.com

M. Diwakar
Department of Computer Science Engineering, Graphic Era Deemed to Be University, Dehradun, India
e-mail: manoj.diwakar@gmail.com

M. Tiwari
Department of Computer Science & Engineering, Bharati Vidyapeeth College of Engineering, Delhi, India
e-mail: mohit.t.bvcoe@gmail.com

© The Author(s), under exclusive license to Springer Nature Switzerland AG 2025
S. Sambhav et al. (eds.), *Empowering Solutions for Sustainable Future in Science and Technology*, SpringerBriefs in Applied Sciences and Technology,
https://doi.org/10.1007/978-3-031-77837-7_1

showcase that the ANN outperforms the LSTM and decision tree models in terms of predicting anomalies. The accuracy, recall, and F1 score of the ANN demonstrate its proficiency in recognizing health deviations while reducing false positives. This predictive capability holds the potential to revolutionize healthcare procedures by ensuring early identification and appropriate care. The research underscores the significance of incorporating predictive analytics into IoT-enabled healthcare frameworks. The continuous monitoring, proactive forecasting, and remote accessibility of such systems cater to the evolving needs of modern health care.

Keywords IoT health care · Machine learning · Elderly care · Health monitoring system · Patient care technology · Smart health care

1 Introduction

With technological improvements over the last few decades, the healthcare business has swiftly evolved. The Internet of Things (IoT) is a linked network of systems, sensors, and systems that allows for seamless data exchange and communication [1]. The utility of IoT tools in health care in recent years has opened up new possibilities for patient monitoring and care delivery. This paradigm change may have an impact on the delivery and management of healthcare services, particularly in situations requiring continuous monitoring, such as the care of elderly patients and severe cases in intensive care units (ICUs) [2, 3].

Many people are interested in integrating IoT into healthcare systems because of its potential to increase patient outcomes, lower healthcare prices, and optimize resource allocation [4, 5]. The current study, which focuses on the use of IoT in continuous health monitoring with an emphasis on elderly patients, adds to the expanding body of research. As the world population ages and the frequency of chronic illnesses rises, there is a rising demand for improved monitoring technologies that may detect health abnormalities early and give prompt treatment. The examiner underlines the importance of IoT-enabled progresses in getting together this need, opening the entryway for a more proactive and customized approach to health care [6].

The ampleness of IoT-based prosperity checking is subordinate on the smooth integration of some specialized components, such as sensors, machine learning calculations, and cloud computing. These components work together to form an add-up to framework for collecting, putting absent, and analyzing health-related data [7, 8]. This sort of advancement has the potential to move forward the accuracy and ampleness of prosperity abnormality recognizing confirmation by allowing therapeutic experts to recognize assortments from design prosperity data in real time. This usefulness is especially imperative in broad settings, where more distant watching is required to decrease the chance of introduction for both patients and remedial specialists [9, 10].

As demonstrated in this study, using machine learning approaches such as long short-term memory (LSTM), decision tree, and artificial neural network (ANN)

enhances the potential of IoT-enabled health monitoring. These algorithms are capable of deciphering complicated patterns in sensor data and accurately predicting health problems. The comparison of these models in terms of predictive performance gives useful insights into their relative strengths and limits, illuminating which model holds the most potential for real-world healthcare applications [11–13].

Aside from technological developments, this study emphasizes the need of predictive analytics in healthcare settings. Because of its capacity to change the healthcare environment, the notion of harnessing data-driven insights to forecast health trends and anomalies has gained appeal [12]. By allowing for the early detection of possible health concerns, healthcare practitioners may intervene before a disease worsens, resulting in better patient outcomes and a lower strain on healthcare systems. Furthermore, the study underlines the significance of remote accessibility, which is consistent with the growing trend of telemedicine and remote patient care. IoT-based systems provide continuous monitoring and proactive intervention, particularly in settings where in-person healthcare services may be difficult to provide.

2   Architecture of IoT in the Health Sector

This study investigates how Internet of Things technology may be used in the healthcare industry, with an emphasis on monitoring the health of elderly patients. The Internet of Things has provided several opportunities in a range of industries, including health care. Adoption of IoT-based solutions in health care has shown promising potential, particularly in situations where continuous health monitoring is required, such as in the case of elderly patients in an intensive care unit (ICU) or those receiving home treatment, especially during pandemics where minimizing in-person interactions is critical.

The primary purpose of this project is to develop a full IoT framework that uses the capabilities of many sensors' data in real time. The architecture of the research is shown in Fig. 1. Temperature and pulse sensors, for example, can be used to capture critical physiological data. These sensors serve as the initial line of data collection, providing a consistent stream of data on patient health.

To efficiently obtain, evaluate, and disseminate this data, the study used a multi-tier approach. Arduino microcontrollers are utilized extensively to interface with the numerous sensors. These microcontrollers serve as data aggregators, collecting raw sensor data and converting it into a format suitable for further analysis. The Arduino is a popular choice because of its versatility and interoperability with a wide range of sensor types, since it acts as an ideal bridge between sensors and more powerful IoT framework components.

ESP8266 microcontrollers become increasingly significant as one progresses along the architecture. They are in charge of collecting data from Arduino that has already been processed and transmitting it to the cloud. The ESP8266 modules are well-known for their compact form factor and integrated Wi-Fi capabilities, and

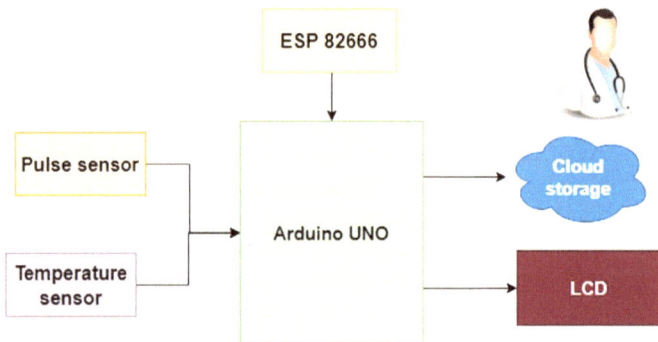

Fig. 1 Architecture of the proposed research

they enable a seamless wireless connection. This is especially valuable in health-care settings where mobility, ease of installation, and continuous data sharing are required.

The system's beating heart is the cloud computing component. It serves as a center for storage, data analysis, and remote access. The information transmitted from the ESP8266 modules is stored in the cloud using secure communication protocols. In addition to serving as a patient health archive, this data collection also lays the foundation for further research. Real-time anomaly, trend, and potential health concern identification are made possible by the use of powerful analytics and machine learning technology.

3 Working of the Proposed Research

With a primary focus on continuous health monitoring for elderly patients, particularly those in intensive care units or receiving home treatment during pandemics, the proposed study is built on a meticulously designed framework that seamlessly integrates Internet of Things technology into the healthcare sector. Real-time health monitoring and remote access are successfully implemented in this innovative system, thanks to a network of interconnected operations.

The usage of Arduino microcontrollers follows as the next step in the process. These microcontrollers serve as a connection point between sensors and more advanced IoT architectural elements. They collect sensor data, process it, and then transmit it to ESP8266 modules. The Arduino is an excellent candidate for this job due to its scalability and compatibility with a variety of sensors.

ESP8266 devices with wireless communication capabilities take center stage in the architecture. They receive pre-processed data from Arduino microcontrollers and communicate with the cloud computing platform over a secure wireless connection.

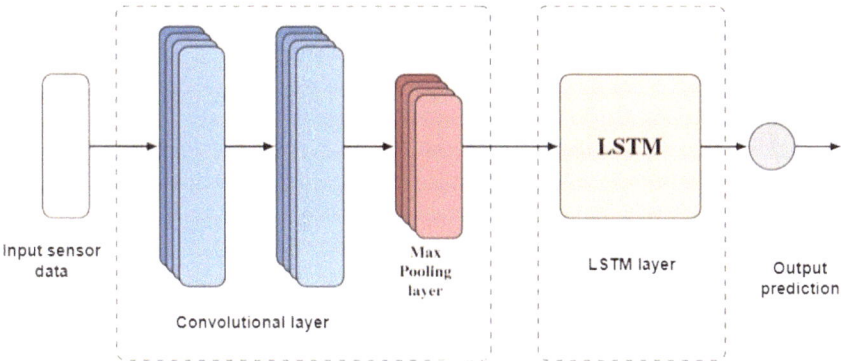

Fig. 2 LSTM model

This continuous data transfer guarantees that health data is always up to date and accessible for analysis and remote access.

3.1 Machine Learning Algorithm

In the context of health monitoring, long short-term memory can detect patterns that suggest imminent health crises by capturing temporal relationships within physiological data. The architecture of the LSTM is shown in Fig. 2. Because of its intrinsic capacity to learn from previous data while taking into account the sequence of events, LSTM is particularly good at forecasting complicated physiological patterns, making it a critical tool for early detection.

The decision tree method, on the other hand, is highly regarded for its ease of use and interpretability. This method splits data depending on qualities, resulting in a decision tree-like structure. Decision trees may be used in health care to analyze the relative relevance of various patient metrics, assisting in the identification of crucial indications for health worsening.

4 Result and Discussion

The suggested use of IoT technology in health care is a critical step toward changing patient care and monitoring. A complete testing phase including a cohort of 50 people was done to systematically analyze the usefulness and dependability of this unique framework. This heterogeneous group reflects a range of senior people, with 38 participants classified as healthy and the remaining 12 participants classified as having various health issues.

The 38 healthy older volunteers provide as a benchmark for typical health markers, allowing any variations to be discovered. The data from this group is used to train machine learning algorithms to spot typical patterns and forecast normal health outcomes. The subset of 12 volunteers with diverse health issues, on the other hand, adds a significant dimension to the testing phase. Figure 3 depicts temperature measurements in degrees Celsius and pulse rate data in beats per minute (bpm). This set of example data is intended to imitate the readings obtained from the sensors integrated into the proposed IoT healthcare system.

Each row in Table 1 corresponds to a sample sensor reading. The physiological data obtained is represented by the columns temperature and pulse rate. The LSTM prediction, decision tree prediction, and ANN prediction columns provide the results of each model's analysis based on the sensor data presented. The LSTM model discovered irregularities in samples 3 and 6, where both temperature and pulse rate had higher values, according to Table 1. The decision tree also accurately discovered the anomalies in samples 3 and 6. However, the ANN performed better, correctly anticipating the anomalies in all cases (samples 3, 6, and 8) when they occurred.

In Table 2, the accuracy column reflects the percentage of accurate projections presented by each model. The measurements show that the LSTM and decision tree models both have an accuracy of 70%. The ANN performs better than both, obtaining a 90% accuracy. This increased accuracy demonstrates how much better the ANN is at making predictions.

Precision is crucial in the healthcare industry since false positives might result in pointless treatments. The ANN has a remarkable accuracy of 100%, meaning that it is very likely to be accurate when it predicts an anomaly. The LSTM and decision tree models, in comparison, have accuracy that is 50% and 66.67%, respectively. Recall is similarly important since it gauges how well the model can recognize real anomalies. The decision tree has a score of 1, while the ANN and LSTM both have a recall score of 66.67%.

Fig. 3 Sensor readings

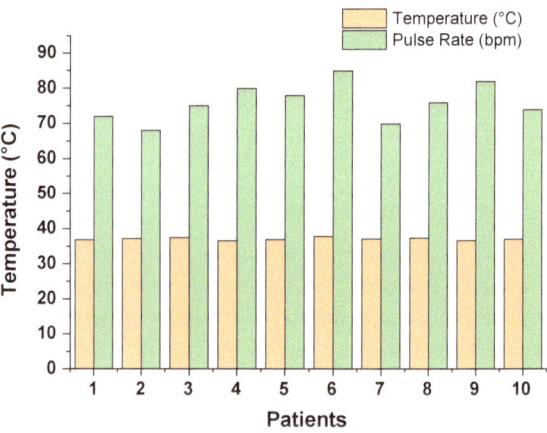

Table 1 Prediction from the each model

Sample	Temperature (°C)	Pulse rate (bpm)	LSTM prediction	Decision tree prediction	ANN prediction
1	36.8	72	0	0	0
2	37.2	68	0	0	0
3	37.5	75	1	1	1
4	36.6	80	0	0	0
5	36.9	78	0	0	0
6	37.8	85	1	1	1
7	37.1	70	0	0	0
8	37.4	76	1	0	1
9	36.7	82	0	0	0

Table 2 Performance evaluation score

Model	Accuracy (%)	Precision (%)	Recall (%)	F1 score (%)
LSTM	70	50	66.67	57.14
Decision tree	70	66.67	66.67	66.67
ANN	90	100	66.67	80

The ANN receives the highest F1 score (80%), demonstrating its ability to accurately forecast both the existence and absence of health abnormalities. The LSTM trails behind with an F1 score of 57.14%, while the decision tree comes in second with a score of 66.67%.

5 Conclusion

Finally, this research examines the mixing of Internet of Things technology into the healthcare business, with a focus on continuous health monitoring for elderly patients. By combining a network of sensors, machine learning algorithms, and cloud computing, the proposed IoT-based healthcare system has demonstrated its ability to change patient care and monitoring operations. The testing phase of the study, which included both healthy and sick older persons, offered critical insights regarding the framework's efficacy. Because of the diverse dataset, machine learning algorithms such as LSTM, decision tree, and ANN may be trained and evaluated. The findings emphasized the ANN's exceptional accuracy in predicting health issues, firmly establishing it as the most effective model for early detection and care. By recognizing deviations from baseline health conditions and immediately informing healthcare practitioners, the system has the capability to enhance patient consequences and reduce the impact of health emergencies. As the healthcare industry advances, the IoT-based healthcare framework has immense promise.

References

1. S.M. Rajagopal, M. Supriya, R. Buyya, FedSDM: federated learning based smart decision making module for ECG data in IoT integrated Edge–Fog–Cloud computing environments. Internet Things 100784 (2023)
2. A.J. Perez, F. Siddiqui, S. Zeadally, D. Lane, A review of IoT systems to enable independence for the elderly and disabled individuals. Internet Things 100653 (2023)
3. A.H. Sodhro, A.I. Awad, J. van de Beek, G. Nikolakopoulos, Intelligent authentication of 5G healthcare devices: a survey. Internet Things 100610 (2022)
4. N. Singh, S.P. Sasirekha, A. Dhakne, B.V.S. Thrinath, D. Ramya, R. Thiagarajan, IOT enabled hybrid model with learning ability for E-health care systems. Meas.: Sens.S. 24, 100567 (2022)
5. S. Kumar, S.K. Pathak, A comprehensive study of XSS attack and the digital forensic models to gather the evidence. ECS Trans. **107**, 7153 (2022)
6. S. Mall, Heart diagnosis using deep neural network, in *Paper presented at the 3rd International Conference on Computational Intelligence and Knowledge Economy (ICCIKE)*, Amity University, (Dubai, 2023), pp. 9–10
7. A. Sharan, M. Saini, Term co-occurrence and context window-based combined approach for query expansion with the semantic notion of terms. Int. J. Web Sci. **3**(1), 32–57 (2017)
8. P. Singhal, S. Gupta, An integrated approach for analysis of electronic health records using blockchain and deep learning. Recent. Adv. Comput. Sci. Commun. **16**(9), 1–10 (2023)
9. M. Sajid, M.S. Jawed, S. Abidin, M. Shahid, S. Ahamad, Capacitated vehicle routing problem using algebraic harris hawks optimization algorithm, in *Intelligent Techniques for Cyber-Physical Systems*, ed. M. Sajid, A. Kumar Sagar, J. Singh, O.I. Khalaf, M. Prasad (CRC Press, Florida, 2023)
10. S. Sellamuthu, S.A. Vaddadi, S. Venkata, H. Petwal, R. Hosur, V. Mandala, AI-based recommendation model for effective decision to maximise ROI. Soft Comput. 1–10 (2023)
11. C.S. Yadav, M.K. Pradhan, S.M.P. Gangadharan, J.K. Chaudhary, J. Singh, A.A. Khan, M.A. Haq, Multi-class pixel certainty active learning model for classification of land cover classes using hyperspectral imagery. Electronics **11**(17), 27299 (2022)
12. R. Kumar, Lexical co-occurrence and contextual window-based approach with semantic similarity for query expansion. Int. J. Intell. Inf. Technol. **13**(3), 57–78 (2017)
13. M. Prasad, Y.A. Daraghmi, P. Tiwari, P. Yadav, N. Bharill, M. Pratama, A. Saxena, Fuzzy logic hybrid model with semantic filtering approach for pseudo relevance feedback-based query expansion, in *Paper presented at 2017 IEEE symposium series on computational intelligence (SSCI)*, Hawaii, USA, 27–30 November 2017

Edge-Enabled Cloud IoT System for Multi-Disease Health Care: Predictive Approach for Elderly Patients

Amit Kumar Mishra, Rahul Sharma, Jagendra Singh, Shilpi Singh, Manoj Diwakar, and Mohit Tiwari

Abstract The escalating demand for timely and precise medical care, especially for the elderly, necessitates advanced technological interventions. This research delved into an edge-enabled cloud IoT system tailored for multi-disease health care, presenting a novel confluence of edge and cloud computing. Grounded in real-world scenarios, the system employed machine learning algorithms, including logistic regression, decision trees, and support vector machines, to predict heart, kidney, and brain diseases. Through a meticulously designed framework, real-time data processing at the edge was harmoniously integrated with in-depth analysis in the cloud. The results showcased the system's potent capability to predict potential health concerns with significant accuracy. For the elderly demographic, the system transitioned health care from a reactive to a proactive paradigm, with continuous monitoring acting as a pivotal asset. The research serves as a beacon, highlighting the

A. K. Mishra
Department of Computer Science & Engineering, Graphic Era Hill University, Dehradun, India
e-mail: amitmishraddun@gmail.com

R. Sharma
D Y, Patil International University, Pune, India
e-mail: prof.rahuls@gmail.com

J. Singh (✉)
School of Computer Science Engineering & Technology, Bennett University, Greater Noida, India
e-mail: jagendrasngh@gmail.com

S. Singh
Amity School of Engineering and Technology Patna, Amity University, Patna, India
e-mail: ssingh3@ptn.amity.edu

M. Diwakar
Department of Computer Science Engineering, Graphic Era Deemed to Be University, Dehradun, India
e-mail: manoj.diwakar@gmail.com

M. Tiwari
Department of Computer Science & Engineering, Bharati Vidyapeeth College of Engineering, Delhi, India
e-mail: mohit.t.bvcoe@gmail.com

transformative potential of integrating advanced computing paradigms with health care, offering enhanced patient outcomes, especially for vulnerable groups. This study is not merely an exploration but a foundational step toward a future where technology and health care seamlessly intertwine, ensuring holistic well-being.

Keywords Edge-enabled IoT · Cloud computing · Machine learning · Multi-disease prediction · Elderly healthcare

1 Introduction

In the modern world, the rapid proliferation of the Internet of Things (IoT) has profoundly impacted multiple sectors, ushering in an era of unprecedented connectivity and digital transformation. Health care, a sector pivotal to human well-being, has been one of the prominent beneficiaries of this technological revolution [1]. The integration of IoT in health care—often referred to as the Internet of Medical Things (IoMT)—promises a paradigm shift in how care is delivered, monitored, and optimized. With devices ranging from wearable fitness trackers to specialized implantable devices, the capacity to monitor various health metrics in real time has become a tangible reality [2]. The inception of IoMT hasn't merely been about connectivity; it is about the insights these connections offer. Today's world grapples with an array of health challenges, many of which are multifaceted and interconnected. For instance, an aging global population faces the looming threat of chronic diseases, many of which might overlap or have similar symptoms. Heart disease, kidney disorders, and brain-related ailments are among the leading causes of morbidity and mortality among the elderly. Consequently, the ability to predict, prevent, and manage such multi-diseases has never been more critical [3].

While the foundational premise of IoMT is compelling, the sheer volume of data it generates is staggering. Herein lies the significance of edge-enabled cloud IoT systems [4, 5]. Traditional centralized cloud computing models, while robust and scalable, sometimes falter in real-time data processing needs of the healthcare domain, especially when milliseconds can be the difference between life and death. Edge computing emerges as the solution to this quandary. By processing data at the edge—closer to where it is generated—latency is drastically reduced, and the real-time processing of critical health data is enabled. Combining the massive storage and computational power of the cloud with the immediacy of edge computing creates a synergistic model, perfectly attuned to the demands of modern health care [6, 7].

The domain of body sensor networks (BSN) has witnessed exponential growth, with diverse sensor systems making inroads into the medical field. Initially conceptualized for general health and fitness monitoring, BSNs have morphed into sophisticated networks, each tailored for specific medical applications. For instance, early sensors were primarily wearables like smartwatches or fitness bands [8]. But as these networks expand, so does the complexity of managing and processing their data, revealing the need for robust computational paradigms. This complexity segues into the evolving world of distributed edge computing [9, 10].

As the complexity of data grew, so did the necessity for tools capable of deciphering intricate patterns and making predictions. ML algorithms, especially deep learning models, have shown an uncanny ability to identify subtle patterns, often imperceptible to human analysis [11, 12]. Examples abound, from predicting the onset of sepsis in ICU patients based on real-time monitoring to using imaging data to forecast the likelihood of tumor malignancy [13, 14].

2 System Framework

In the labyrinth of healthcare technology, a system's architecture plays a pivotal role in determining its efficacy, scalability, and reliability as shown in Fig. 1. The proposed IoT system framework endeavors to address these dimensions, striving for an architecture that harmonizes data collection, processing, and predictive analytics to better serve elderly patients facing multiple health concerns.

2.1 Overview of the Proposed System

At its core, the proposed system is a symbiotic integration of body sensor networks, cloud computing, and edge computing. BSNs act as the primary data collectors, continuously monitoring various physiological metrics from patients. This data, once collected, doesn't follow a linear path. Instead, its journey is determined by its nature and urgency.

Cloud computing serves as the backbone of this architecture. It offers several essential functions:

- Data Storage: Given the continuous stream of data from multiple BSNs, cloud platforms provide vast storage capacities to archive this information.
- Advanced Analytics: While immediate data processing might occur at the edge, the cloud hosts more resource-intensive machine learning algorithms. These algorithms periodically analyze aggregated data, improving predictive models and offering insights into long-term health trends.
- Accessibility: The cloud ensures that patient data is accessible to authorized personnel from anywhere, be it doctors seeking patient histories, researchers studying broader health patterns, or patients themselves reviewing their health metrics.

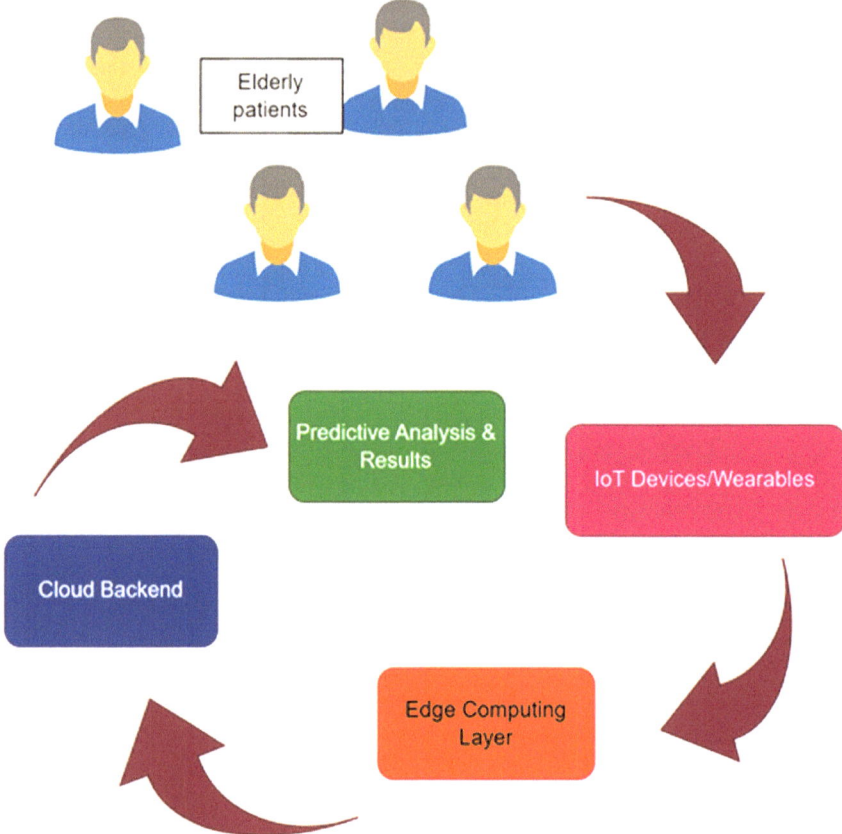

Fig. 1 Proposed algorithm

Edge computing is the system's reflexive arm, acting swiftly when immediacy is paramount. Its functions include

- Real-time Processing: For data that demands immediate attention, such as sudden cardiac irregularities, edge devices process this information in real time, triggering alerts or interventions as needed.
- Local Data Storage: Edge devices can store data locally, reducing the need for continuous data transmission to the cloud. This is particularly valuable in areas with unstable Internet connectivity.
- Offloading Cloud Bandwidth: By processing a portion of data locally, edge computing reduces the data load on cloud servers, ensuring smoother, uninterrupted performance.
- In summation, the proposed IoT system framework is an intricate dance of data flow between sensors, edge devices, and cloud platforms. It aims to harness the

strengths of each component, creating a robust, responsive, and scalable architecture tailored to the unique challenges of multi-disease health care for the elderly.

2.2 Predictive Analysis Using Machine Learning

Harnessing the power of machine learning (ML) for predictive analysis has been a transformative shift in health care. With the deluge of data streaming from various sensors, traditional methods of data analysis fall short in detecting intricate patterns and making accurate predictions. Machine learning, with its data-driven approach, fills this gap. Instead of relying on explicit programming, ML models learn from data, tweaking their internal parameters to optimize predictions. In the context of health care, this translates into a system that can anticipate potential health issues before they manifest. For instance, an ML model might analyze weeks of heart rate data, identify subtle irregularities, and predict an impending cardiac event. The more data the model is exposed to, the better its predictions become, enabling healthcare professionals to intervene proactively and potentially prevent adverse outcomes.

2.3 Signature Analysis of Heart-, Kidney-, and Brain-Related Diseases

In Table 1, for heart diseases, signature analysis might revolve around studying ECG data. Specific patterns like the T-wave inversions or ST-segment elevations can be indicative of conditions like ischemia. Machine learning models, trained on vast datasets of annotated ECG readings, can detect these patterns even when they are subtle, heralding early stages of cardiac conditions. In the realm of kidney diseases, the signatures are often biochemical. For instance, a sudden rise in serum creatinine might indicate acute kidney injury. ML models can be trained to analyze periodic blood test results, identify such signatures, and predict the onset or progression of renal ailments. Brain-related conditions pose a unique challenge due to the complexity of neural activity. However, tools like EEGs provide a window into brain wave patterns. Conditions like epilepsy or even neurodegenerative diseases might manifest in specific EEG signatures.

Table 1 Disease indicators

Organ system	Potential indicators
Heart	Abnormal ECG readings, arrhythmias, unusually prolonged QT intervals, and fluctuations in heart rate variability
Kidney	Elevated creatinine levels, changes in glomerular filtration rate, presence of proteins in urine, and alterations in electrolyte balance
Brain	Unusual EEG patterns, changes in neural response times, and alterations in specific neural waveforms like delta or theta waves

3 Integration with Cloud and Edge Computing

The future of healthcare technology rests heavily on the seamless integration of various components, ensuring that data flows efficiently and interventions are timely. In the landscape of IoT systems, two pivotal components emerge: cloud computing and edge computing.

3.1 Data Collection

One of the critical aspects of the IoT system's deployment in a healthcare setting is the structured collection of data. When we gaze upon the organized numerical values of our sample data collection table, we aren't just viewing numbers; we are observing a narrative of health, disease, and potential prognosis for ten distinct individuals as shown in Fig. 2. Beginning with the pulse rate, measured in beats per minute (bpm), it offers a window into the cardiovascular health of an individual. The average resting adult heart rate ranges 60–100 bpm, and our table illustrates that all ten patients lie comfortably within this bracket. Patients P002 and P007, with pulse rates of 68 and 69 bpm respectively, veer toward the lower end of this spectrum, indicating a potentially more efficient cardiovascular system or a more relaxed state at the time of measurement.

Our dataset demonstrates that all patients lie in the lower echelons of this range, hinting at efficient kidney function. Patient P003, with a urea content of 14 mg/dL, and Patient P008, at 20 mg/dL, represent the span of our dataset. If visualized on a graph, one would observe a cluster of values in the mid-teens to early twenties, reinforcing the narrative of a group with healthy kidney function.

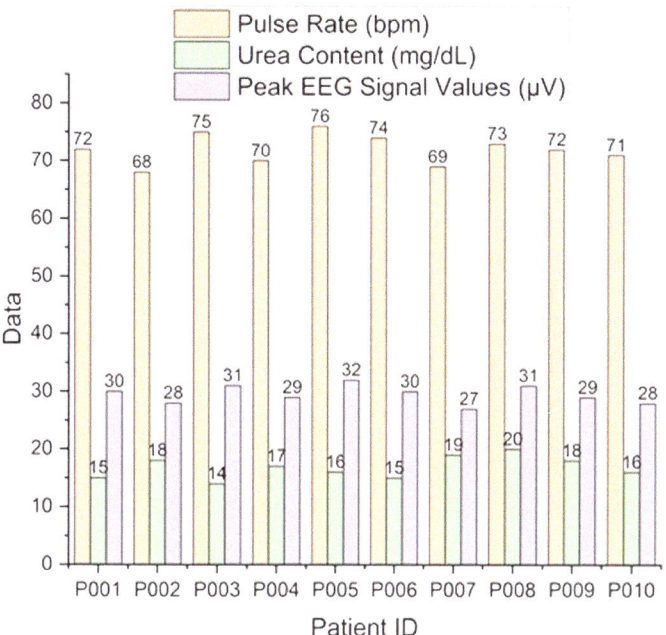

Fig. 2 Data collection

4 Data Analysis

4.1 Logistic Regression

Logistic regression is a statistical method for analyzing datasets in which there are one or more independent variables that determine an outcome. The outcome is typically a binary outcome (1/0, Yes/No, True/False). Let us predict based on.

- A pulse rate above 75 bpm suggests a higher likelihood of heart disease.
- A urea content above 18 mg/dL indicates potential kidney issues.
- Peak EEG signal values above 31 μV suggest possible brain-related concerns.

Analyzing the insightful narratives on each patient's health trajectory, for instance, Patient P005 exhibits a 60% likelihood of developing heart disease, the highest among the cohort, could be attributed to their relatively higher pulse rate, potentially hinting at underlying cardiovascular issues. On the other hand, patient P008 stands out with a 70% likelihood of kidney disease, perhaps due to their elevated urea content. However, it is essential to remember that these predictions, while rooted in data, are probabilistic in nature. A 60% likelihood doesn't definitively indicate disease onset but rather suggests that, out of a similar patient cohort, 60% might develop the condition.

4.2 Decision Tree

Decision trees are a type of supervised machine learning algorithm that is largely used for classification problems. In the medical field, decision trees can be particularly insightful as they mimic human decision-making processes. In analyzing, patient P003 stands out with a predicted "Yes" for both heart and brain diseases. This suggests that based on the data and the decision nodes of our tree, P003 may be at a higher risk and might need more comprehensive medical evaluations. Meanwhile, patients like P002 and P007 are flagged for potential kidney concerns due to their urea content surpassing the defined threshold.

4.3 Support Vector Machines

In essence, support vector machines (SVMs) work by finding a hyperplane that best divides a dataset into classes. In the context of our healthcare dataset, SVMs can be utilized to classify patients based on the likelihood of them developing heart, kidney, or brain diseases.

5 Conclusion

This research journey into the realm of edge-enabled cloud IoT systems for multi-disease health care unveiled the transformative potential of integrating edge and cloud computing, especially in the context of elderly patient care. Leveraging state-of-the-art machine learning algorithms, including logistic regression, decision trees, and support vector machines, the study demonstrated a potent capability to predict heart, kidney, and brain diseases with considerable accuracy. The real-time data processing afforded by edge computing, combined with the holistic data analysis capabilities of cloud systems, crafted a healthcare model that was both immediate and insightful. Notably, for the elderly, a demographic often battling multiple health concerns, the system offered a paradigm shift. Instead of reactive care, the approach championed proactive interventions, with continuous health monitoring becoming an invaluable asset. Yet, as with any pioneering venture, the path forward beckons further refinements, expansions, and patient-centric modifications.

References

1. C.W. Hsu, C.W. Lee, S.C. Hsu, W.C. Huang, Y.P. Hsu, M.J. Chi, Improvement of the identification of seniors at risk scale for predicting adverse health outcomes of elderly patients in the emergency department **68**, 101274 (2023)

2. J.A. Santing, C.L. Van Den Brand, M.J. Panneman, J.S. Asscheman, J. van der Naalt, K. Jellema, Increasing incidence of ED-visits and admissions due to traumatic brain injury among elderly patients in the Netherlands, 2011–2020. Injury 110902 (2023)
3. P. Singhal, S. Gupta, J. Singh, An integrated approach for analysis of electronic health records using blockchain and deep learning. Recent. Adv. Comput. Sci. Commun. **16**(9), 1–10 (2023)
4. M. Sajid, M.S. Jawed, S. Abidin, M. Shahid, S. Ahamad, Capacitated vehicle routing problem using algebraic Harris hawks optimization algorithm. *Intelligent Techniques for Cyber-Physical Systems* (CRC Press, 2022), pp. 183–210
5. S. Sellamuthu, S.A. Vaddadi, S. Venkata, H. Petwal, R. Hosur, V. Mandala, J. Singh, AI-based recommendation model for effective decision to maximise ROI. Soft Comput. 1–10 (2023)
6. S. Mall, Heart diagnosis using deep neural network, in *Paper presented at the 3rd International Conference on Computational Intelligence and Knowledge Economy (ICCIKE)* (Amity University, Dubai, 2023)
7. A. Sharan, M. Saini, Term co-occurrence and context window-based combined approach for query expansion with the semantic notion of terms. Int. J. Web Sci. **3**(1), 32–57 (2017)
8. C.S. Yadav, A. Yadav, H.S. Pattanayak, R. Kumar, A. Khan, Mal-ware analysis in IoT & android systems with defensive mechanism. Electronics **11**(15), 2354 (2022)
9. A. Tyagi, D. Rajpal, A. David, J. Singh, H.K. Thakur, K. Upreti, in *2023 OITS International Conference on Information Technology (OCIT)*, (Raipur, India, 2023)
10. S.E. Bakyarani, N.P. Singh, J. Shekhawat, S. Bhardwaj, S. Chaku, J. Singh, A novel approach on deep reinforcement learning for improved throughput in power-restricted IoT networks, in *Innovations in Electrical and Electronic Engineering* (Springer, Singapore, 2024)
11. S. Agarwal, R. Sharma, M. Tamilselvi, H.M. Sharma, D.P. Sahu, J. Singh, in *2023 International Conference on Computing, Communication, and Intelligent Systems (ICCCIS)* (Springer, Greater Noida, India, 2023)
12. C.S. Yadav, M.K. Pradhan, S.M.P. Gangadharan, J.K. Chaudhary, J. Singh, A.A. Khan, M.A. Haq, Multi-class pixel certainty active learning model for classification of land cover classes using hyperspectral imagery. Electronics **11**(17), 27299 (2022)
13. A. Tyagi, R. Khandelwal, N.A. Shelke, J. Singh, D. Rajpal, I.R. Gaware, Comparitive Analysis of various transfer learning apporaches in deep CNNs for image classification, in *Recent Trends in Image Processing and Pattern Recognition* University in Derby (Springer, Derby, 2024)
14. R. Kumar, Lexical co-occurrence and contextual window-based approach with semantic similarity for query expansion. Int. J. Intell. Inf. Technol. **13**(3), 57–78 (2017)

Technological Dynamics of Digital Democracy: Social Media's Influence on Voters' Intention

Kishore Bhattacharjee, Chetna Priti, Rohit Kumar, and Ajit Kumar

Abstract This study addresses a critical gap in understanding the dynamics of modern electoral campaigning, particularly the influence of social media on voter behavior. By examining the impact of voters' personal norms, social norms, and perceived behavioral control on their voting intentions, with a specific focus on the moderating role of social media political campaigns, this research sheds light on the intricate interplay between technology and politics. The study utilizes a two-stage structural equation modeling (SEM) approach and a sample size of 318 respondents. It reveals that social norms, personal norms, and perceived behavioral control significantly shape voting intentions, underscoring the central role of social media in shaping, and reinforcing these psychological factors. These findings have critical implications for political parties and candidates in the digital age. By strategically aligning their messaging with these psychological determinants and leveraging the influence of digital technology such as social media platforms, political actors can actively mold the social norms that impact voter decisions.

Keywords Social media campaign · Personal norms · Social norms · Perceived behavioral control · Political marketing

K. Bhattacharjee (✉) · C. Priti · R. Kumar
Amity Business School, Amity University Patna, Patna, Bihar, India
e-mail: kbhattacharjee@ptn.amity.edu

C. Priti
e-mail: cpriti@ptn.amity.edu

R. Kumar
e-mail: rkumar1@ptn.amity.edu

A. Kumar
Amity College of Commerce and Finance, Amity University Patna, Patna, Bihar, India
e-mail: akumar@ptn.amity.edu

© The Author(s), under exclusive license to Springer Nature Switzerland AG 2025
S. Sambhav et al. (eds.), *Empowering Solutions for Sustainable Future in Science and Technology*, SpringerBriefs in Applied Sciences and Technology,
https://doi.org/10.1007/978-3-031-77837-7_3

19

1 Introduction

The global landscape of technological progress, particularly in the realm of information and communication technology, has revolutionized the conduct of electoral campaigns. This dynamic shift has given rise to a new era of digital democracy, where the boundaries between candidates and voters are blurred and the potential for political influence is more democratized than ever before. Political parties worldwide are leveraging their presence on social media to convey their message moving away from traditional print and electronic media with the aim to reach, inform, and persuade voters digitally [1].

In recent years, India has also witnessed a remarkable surge in its online population, and Indian political parties have recognized the potential of these platforms for reaching a vast and diverse audience using social media. Numerous studies have delved into the role of social media campaigning in assessing various aspects of voters' behavior, including their intentions, attitudes, emotions, ideology, and loyalty to political parties [2]. The quality of promotional content on social media has been found to be critical in influencing voter engagement, trust, and the overall impact of political marketing efforts [3]. The quality of promotional content on social media plays a pivotal role in shaping voter attitudes and behaviors.

Nevertheless, the current body of literature lacks a comprehensive framework for scrutinizing the development of voting intentions (VI) influenced by social media promotion. Consequently, there is an evident research gap necessitating a focused effort to gain a thorough understanding of how political campaigning through social media can moderate the voters' personal and social norms as well as their perceived behavioral control over their intention to vote. This study aims to investigate the influence of voters' personal norms (PN), social norms (SN), and perceived behavioral control (PBC) on their voting intention (VI) while examining the moderating role of social media political campaigning within these relationships.

Theoretical Framework

The theoretical framework for this study is built upon the theory of planned behavior (TPB) and the elaboration likelihood model (ELM). The TPB provides a comprehensive framework for understanding and predicting human behavior, making it applicable for measuring voting intentions [4], while the ELM is often used to understand how persuasive messages are processed and how certain factors, like social media campaigning, can influence the impact of these messages [5]. The ELM allows for the examination of interaction effects, where the impact of PN, SN, and PCB on VI may vary based on the effectiveness of social media campaigning.

Hypothesis Development
Personal Norms (PN) and Voting Intentions (VI)

The concept of PN pertains to an individual's moral sense of obligation, either to engage in or refrain from specific actions [6]. Given that a voter's perception of political messages on social media is clearly impacted by their internalized moral responsibilities, it has been considered advantageous to incorporate PN as a factor in current research for predicting a voter's position on political messages presented on social media. The inclusion of PN can enhance the predictive effectiveness of the existing TPB model in explaining a voter's attitude toward political advertising through social media. Consequently, we posit the following hypothesis:

H1: Voters' PN significantly influence their VI.

Social Norms (SN) and Voting Intentions (VI)

Subjective norms, often considered social norms, encompass the influence of social pressure on an individual's decision-making regarding their participation in or avoidance of a particular behavior [4]. Scholars have consistently described SN as potent social pressures that significantly impact the intentions of voters. Numerous studies have delved into the direct impact of subjective norms, which are essentially SN, on behavioral intentions [7]. So, the hypothesis framed in the context is

H2: Voters' SN significantly influence their VI.

Perceived Behavioral Control (PBC) and Voting Intentions (VI)

Perceived behavioral control pertains to an individual's perception of the ease or difficulty associated with performing a particular behavior [4]. Previous research has emphasized the indirect influence of perceived behavioral control on behavioral intentions. However, contemporary scholars have put forth the notion that perceived behavioral control also directly impacts behavioral responses [7]. Consequently, in the context of the current study, we posit the following hypothesis

H3: Voters' PCB significantly influences their VI.

Social Media Political Campaigning (SMPC) as a Moderator

PN are influential factors in shaping political behavior, and social media platforms provide a space for these norms to be both expressed and challenged. Similarly, SN plays a significant role in shaping voter behavior and engagement within the political context. When voters perceive a strong social norm endorsing political engagement, they are more likely to participate in elections.

PCB is another crucial factor that significantly influences political participation. Voters with a high level of perceived behavioral control are more likely to participate in the electoral process by registering, voting, or engaging in campaign activities [8]. High levels of perceived behavioral control enhance individuals' motivation to participate in politics as they feel more capable of taking action [4]. Social media platforms contribute to this by offering easy access to political information and resources.

Moreover, it is evident from the literature that political campaigning through social media serves as a critical factor in affecting PN, SN, PCB, information exposure, candidate messaging, normative influence, etc. Social media's role as an intermediary is crucial in shaping the dynamics of modern political campaigns [8]. Based on these conceptualizations, the following hypotheses and the conceptual model (Fig. 1) have been proposed:

H4a: SMPC significantly moderates the relationship between voters' PN and VI.
H4b: SMPC significantly moderates the relationship between voters' SN and VI.
H4c: SMPC significantly moderates the relationship between voters' PBC and VI.

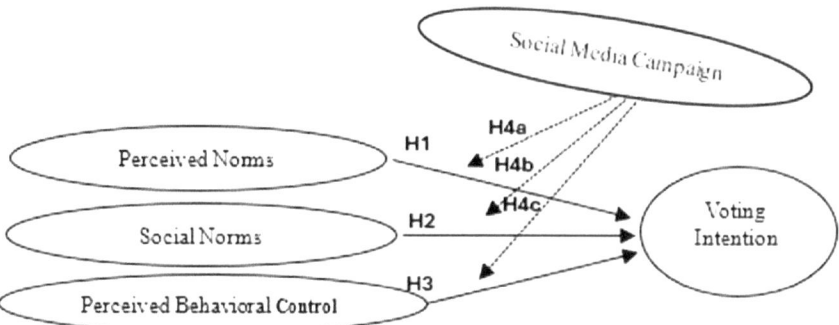

Fig. 1 Conceptual model. Doted lines (………) showing moderation effect. *Source* Author's compilation

2 Methodology

Data and Sample

This study employed a cross-sectional research design to analyze individuals' voting intentions. An online, structured questionnaire was administered to respondents using social media posts and registered e-mails. Since the purpose of this study was to examine the individual's voting intentions, only adults were approached.

318 responses were received in the period of 3 months (January 2023–March 2023). The collected data were first examined for any outliers using Cook's distance approach. Any response having a value larger than one to identify outliers was removed from the sample. 12 responses were rejected because of the outcome. Moreover, skewness and kurtosis indices were used to determine the normality of the data. Since all of the constructs' values for skewness and kurtosis were below three and ten, respectively, the data distribution was regarded as normal. Finally, analysis of the proposed model was performed on 306 responses. Among the 306 samples, 64% were male (n = 196) and 36% were female (n = 110).

Measures, Face Value, and Analytical Tools

All the measuring items for PN, SN, PBC, and VI came from related literature, with only a few words modified to make them easier to understand in the context of voting decisions. Items for each construct were evaluated by respondents on a 5-point Likert scale ranging from 1 ("strongly disagree") to 5 ("strongly agree"). The constructs of the questionnaire have been framed using existing scales, i.e., perceived behavioral control and SN [11], PN [12], social media campaigning [13], and intention toward voting [14]. A pilot study was conducted prior to the final data collection by distributing 35 questionnaires to voters.

The data were then analyzed using a two-stage method of structural equation modeling (SEM). The SEM is used to estimate separate yet interconnected multiple regression equations.

3 Results and Discussion

Measurement Model: Reliability and Validity

A confirmatory factor analysis (CFA) was performed to assess the validity and reliability of the constructs used in the model. For the initial screening of the model, various indicators were accessed and found to fit ($\chi2 = 317.25$; $\chi2$ /df. $= 1.890$; CFI $= 0.940$; GFI $= 0.901$; IFI $= 0.940$; TLI $= 0.931$; and RMSEA $= 0.056$). Composite reliability (CR) was examined to assess scale reliability. The CR values ranged from 0.859 to 0.962, indicating good consistency. Furthermore, three parameters, namely composite reliability (CR), factor loadings, and average variance extracted (AVE), were used to assess convergent and discriminant validity (AVE). The standardized factor loadings of all items were greater than 0.6. The AVE values, which ranged from 0.641 to 0.760, were also higher than the acceptable limit of 0.5. The CR values also exceeded the acceptable limit of 0.6, indicating that the multiple indicators were internally consistent. Furthermore, the square root of AVE was greater than the correlation between each construct, indicating adequate discrimination among the constructs. The results showed that the proposed conceptual model had good validity (both convergent and discriminant) and reliability.

Structural Model: A Path Analysis

Goodness-of-fit statistics were analyzed to assess the overall predictive power of the model. The hypothesized model ($\chi 2 = 306.33$, $\chi 2$ /df $= 1.928$, GFI $= 0.915$, AGFI $= 0.882$, NFI $= 0.943$, TLI $= 0.964$, CFI $= 0.972$, and IFI $= 0.972$) reasonably fits the data. Moreover, the RMSEA value (0.068) was less than the 0.08 suggested guideline. Even though the values for AGFI do not exceed 0.9 (the threshold value), they still met the requirement, which shows the model is acceptable if the value is above 0.8. The analysis of the proposed model (presented in Table 1) indicates that PN ($\beta = 0.193$, p $= 0.046$, t $= 1.999$), SN ($\beta = 0.325$, p $= 0.00$, t $= 3.494$), and perceived behavioral control ($\beta = 0.127$, p $= 0.048$, t $= 2.005$) have a significant direct relationship with the voting intentions of Indian voters. Hence, these results support H1, H2, and H3. The overall predictability of the model ($R2$) was 0.37 (Fig. 2).

Moderation Analysis

To conduct moderation analysis in AMOS, the standardized score for both independent variables and dependent variables was calculated. In the next step, interaction variables were computed. The independent variables, dependent variables, and interaction variables were plugged into AMOS to observe the moderating role played by

Table 1 Result of structural model and hypothesis testing

Hypothesis	Path	Coefficient (β)	Critical ratio (CR)	P-value	Result
H1	PN--- > VI	0.193	2.047	0.046	Supported
H2	SN--- > VI	0.325	3.272	***	Supported
H3	PBC--- > VI	0.127	1.999	0.048	Supported

Source AMOS result

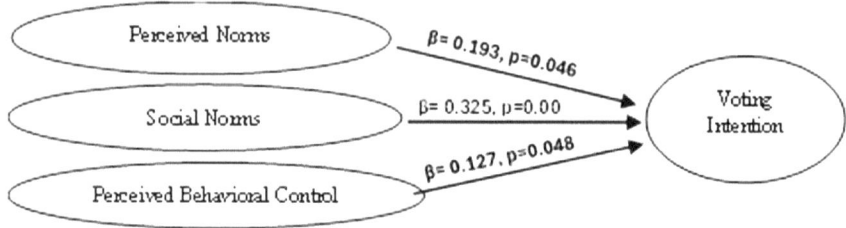

Fig. 2 Result: path analysis. *Source* AMOS output

social media campaigns in the relationship between PN and voting intention, SN and voting intention, and perceived behavioral control and voting intention. The critical ratio (Table 2) is used to report whether the relationships are impacted by respondents' gender or not. The interaction effect of PN and SMC, SN and SMC, and PBC and SMC on VI was found to be significant.

Discussion and Implications

The analysis findings underscore the crucial impact of social media on voting intentions, with a focus on the influence of SN, PN, and PBC [4]. It becomes evident that social and personal norms play significant roles in shaping VI, indicating that social media actively contributes to the formation and reinforcement of these norms within online social circles [9]. This highlights the broader influence of social media, extending beyond information dissemination to actively shaping voters' psychological factors.

The discussion emphasizes the strategic use of social media by political parties to construct and reinforce SN, PN, and perceived behavioral control in favor of their agendas [4, 9]. The literature substantiates the intricate interplay between social media and these psychological determinants in influencing voting intentions and political behavior. This reinforces the notion that political parties can enhance their ability to influence voters by aligning messaging with key psychological factors [4, 9].

The analysis introduces the concept of social media campaigns as moderators in the relationship between psychological determinants and VI. SMPC are shown to significantly moderate the influence of PN, SN, and PBC on VI. This presents an

Table 2 Moderation analysis

Paths	Mod. estimates			Result
	Estimate	P-value	Critical ratio	Comparison result
(H4a): VI<---PN*SMC	0.366	***	2.449***	Supported
(H4b): VI<---SN*SMC	0.426	***	4.298***	Supported
(H4c): VI<---PBC*SMC	0.268	***	1.999***	Supported

*Note *** significantly different from zero at the 0.001 level (two-tailed). Source AMOS result*

opportunity for political actors to strategically use social media to actively shape these psychological determinants in the minds of voters.

To maximize the effectiveness of social media campaigns, political parties are advised to create content that resonates with the personal and social norms of their target audience [10]. Engaging narratives and relatable stories should be central to these campaigns, fostering a personal connection with voters and reinforcing positive behaviors and political engagement. Tailoring messaging to different segments of the population and encouraging meaningful interactions with the audience are also pivotal strategies [10].

Furthermore, the study bridges the theory of planned behavior [4] and the elaboration likelihood model [5], providing a comprehensive framework for understanding the psychological determinants of voting intentions. This integration enriches the literature by combining cognitive components and processing routes in the context of political behavior.

4 Conclusion

In conclusion, the strategic utilization of technology in shaping messages and content allows political entities to harness the influential power of social media [9]. This implies that social media can be a powerful technological tool for political parties to not only connect with voters but also actively mold the SN that influence their voting intentions.

References

1. M. Mannevuo, Uneasy self-promotion and tactics of patience: Finnish MPs ambivalent feelings about personalised politics on social media. Int. J. Cult. Stud. **26**(1), 104–119 (2023). https://doi.org/10.1177/13678779221120028
2. J. Liaukonytė, A. Tuchman, X. Zhu, Frontiers: spilling the beans on political consumerism: do social media boycotts and buycotts translate to real sales impact? Mark. Sci. **42**(1), 11–25 (2023). https://doi.org/10.1287/mksc.2022.1386
3. D.V. Dimitrova, J. Matthes, Social media in political campaigning around the world: theoretical and methodological challenges. J. Mass Commun. Q. **95**(2), 333–342 (2018). https://doi.org/10.1177/1077699018770437
4. I. Ajzen, The theory of planned behavior. Organ. Behav. Hum. Decis. Process. **50**(2), 179–211 (1991). https://doi.org/10.1016/0749-5978(91)90020-T
5. R.E. Petty, J.T. Cacioppo, *Communication and persuasion: central and peripheral routes to attitude change* (Springer-Verlag, New York, 1986)
6. H. Han, Consumer behavior and environmental sustainability in tourism and hospitality: a review of theories, concepts, and latest research. J. Sustain. Tour. **29**(7), 1021–1042 (2021). https://doi.org/10.1080/09669582.2021.1903019
7. C.D. Duong, Using a unified model of TPB, NAM and SOBC to understand students energy-saving behaviors: moderation role of group-level factors and media publicity. Int. J. Energy Sect. Manag. **18**(1), 71–93 (2024). https://doi.org/10.1108/IJESM-09-2022-0017

8. A.O. Larsson, S. Agha, The behavioral immune system as moderator of political behavior: an evaluation of its structure and effects. Evol. Hum. Behav. **39**(4), 327–334 (2018)
9. R.B. Cialdini, R. B. Cialdini, *Influence: the Psychology of Persuasion*, vol. 55 (New York, Collins, 2007), p. 339
10. E. Ferrara, Z. Yang, Measuring emotional contagion in social media. PLOS One **10**(11), e0142390 (2015). https://doi.org/10.1371/journal.pone.0142390
11. S. Taylor, P. Todd, Decomposition and crossover effects in the theory of planned behavior: a study of consumer adoption intentions. Int. J. Res. Mark. **12**(2), 137–155 (1995). https://doi.org/10.1016/0167-8116(94)00019-K
12. M.C. Onwezen, G. Antonides, J. Bartels, The norm activation model: an exploration of the functions of anticipated pride and guilt in pro-environmental behaviour. J. Econ. Psychol. **39**, 141–153 (2013). https://doi.org/10.1016/j.joep.2013.07.005
13. J.C. Athapaththu, K.M.S.D. Kulathunga, Factors affecting online purchase intention: effects of technology and social commerce. Int. Bus. Re. **18**(10), 111–128 (2018). https://doi.org/10.5539/ibr.v11n10p111
14. C.S. Wee, M.S.B.M. Ariff, N. Zakuan, M.N.M. Tajudin, K. Ismail, N. Ishak, Consumers perception, purchase intention and actual purchase behavior of organic food products. Rev. integr. bus. econ. Res. **3**(2), 378 (2014)

A Study on Digital Payment Systems in India with Reference to Global Trends

Kanaka Durga Hanumanthu, Ruchi Shukla, Sailaja Nimmagadda, Shireesha Manchem, and Madhuri Ananthaneni

Abstract Payment systems are evolving continuously, and we try to reach them as per customer requirements. The banking sector is updating itself from the point of view of new service generation by adopting new technologies. Technological upgrading leads to fintech in the banking sector. Through fintech, banks are providing digital payments, neo-banking, etc. With one click, all services are in the footsteps of consumers. Digital payments consist of digital commerce, mobile POS payments, and digital remittances. There is rapid growth in the Indian fintech industry, and, as per estimations, it will reach 150 billion dollars by 2025. According to global statistics, India has the world's third largest fintech industry. So the review and analytical paper focused on the growth and potential market for digital payments in India by focusing on global figures. Global and Asian digital payment quantitative data were compiled and conclusions were drawn. Worldwide and Asian digital payments were correlated with the Indian growth rate, and there is a positive correlation between Indian statistics and global figures too, which suggests that there is a potential and significant market for digital payments. Undoubtedly, the future is for digital technology innovations in the financial sector, and the new era will definitely create a tremendous market for digital payments.

Keywords Digital payments · Technology · Fintech · Growth · Worldwide · Asia

K. D. Hanumanthu (✉) · R. Shukla · S. Nimmagadda
VR Siddhartha Engineering College, Vijayawada, India
e-mail: ykanakadurga@kluniversity.in

R. Shukla
e-mail: ruchis.hs19.phd@nitp.ac.in

S. Nimmagadda
e-mail: nsailaja@vrsiddhartha.ac.in

S. Manchem · M. Ananthaneni
Department of MBA, Andhra Loyola College, Vijayawada, India
e-mail: shireeshabathina@gmail.com

M. Ananthaneni
e-mail: ananthaneimadhuri@gmail.com

1 Introduction

1.1 Digital Payments

The Indian banking sector is well-organized and well-regulated in terms of capital and performance. Payment systems are evolving continuously, and we try to reach them as per customer requirements. Technological upgrading leads to fintech in the banking sector. Through fintech, banks are providing digital payments, neo-banking, etc.

The purpose of the study is to understand facts about digital payments by comparing Indian statistics with global figures, with special reference to Asia (Fig. 1).

2 Literature Review

[1] The paper exhibits trends and shows how people started using digital payments after demonetization in India. Demonetization in India leads to cashless transactions by using electronic modes as well as the Internet. It led to new technological innovation in the fintech era in 2017. [2] A vastly populated country like India needs a sustainable and continuous upgrade of technology in payments that are useful for those who should offer convenient, easy, and secured transactions that allow overcoming the difficulties in the conversion of blocked funds. [3] Digital payments have huge potential due to ever-changing trends in technology as well as society. The literature completely focused on the usage of digital and mobile payment systems. [4] This study gave complete details and an evaluation of the Indian payment application Paytm. It has had its own glory in the market since its inception and has

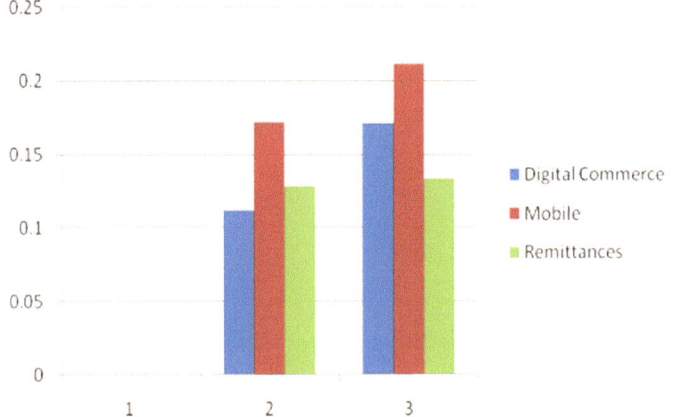

Fig. 1 Growth of digital payments

analyzed how it reaches customers by developing the infrastructure that is needed. [5] A digital payment system leads to enhanced cost efficiency in the banking sector. Observing different factors and adopting technological changes lead to efficiency in transactions on cost matters.

From the above reviews and observations, it is clear that there is a need to analyze present trend in digital payments in quantitative terms, which is a gap in research work. With this brief note, the following objectives have been framed for the present study:

a. To study the trends in the digital payment system in India by comparing global figures.
b. To study the significant growth in digital payment system in India.
c. To study the potential of digital payment systems in India.

3 Research Methodology

The present study is completely focused on secondary data sources collected from different websites, RBI annual reports, different research studies, research articles published in journals, and digital payments from different banks. The data complied with global statistics and was interpreted for the purpose of the study.

Quantitative terms of data were analyzed by using the correlation between the world's growth rate and that of India. Simple percentages were used to analyze the data and draw conclusions.

Hypotheses H0: There is a significant market for digital payments in India.
H1: The Indian digital payment system has more potential to grow.

4 Results and Discussion

4.1 Digital Payments in Europe

In the last three years, digital payments in Europe have increased by up to 30%, reaching 700 million. Mobile wallets such as Apple Pay, Google Pay, Samsung Pay, and P2P such as Venmo and PayPal, etc. are the applications that are widely used on mobile devices for digital payments directly. Mobile banking allows for account-to-account transfers, online or digital transfers to pay regular bills, online individual transfers from account to account, and online purchases. P2P payment platforms, such as Venmo and PayPal, enable individuals to send money to each other instantly via an app. These platforms are often used to split bills, pay friends, or transfer money to family members, whereas in Russia, YooMoney and QIWI are widely used payment applications; in the Netherlands, it is iDEAL; and in Germany,

PayPal. Apart from all these, in the UK, credit cards are still the most preferred means of payment. Overall, 23% of online transactions are with credit cards; the remaining are through e-banking and e-wallets.

4.2 Digital Payments in the USA

Digital payments in the United States of America consist of digital commerce, mobile POS payments, and digital remittances. Digital commerce is related to all online shopping transactions and payments done through credit cards and debit cards by using different Web sources and e-wallets. Different activities in this zone are e-services, travel and tourism, digital media, and digital health. Transaction values are generated via a wide range of activities such as online retail sales, and purchase of services on online market places (including eservices, travel and tourism, digital media, and digital health). The most widespread use of mobile POS systems in the context of digital payments is due to their ability to conduct financial transactions in a flexible and easy manner. Smart phones, QR codes, mobile wallets, and host card emulation are used in this instance to transfer data for the payments. Digital remittances allow non-residents to send money across international borders to residents. MarketWise, Remitly, and Xoom are three commonly used form systems for these payments.

4.3 Digital Payments in India

In India, approximately 114 billion digital payments were registered in the financial year 2022–23 (17 billion in 2021–22). This is a noticeable improvement when compared with the last three years. Digital payments in India included huge interbank payments such as real-time gross settlements (RTGS), national electronic funds transfers (NEFT), and the use of debit and credit cards. Mobile applications like Unified Payments Interface (UPI), which is the most commonly used interface in India, along with global apps like Google Pay and Amazon Pay. The most commonly used domestic fintech start up is the Phone Pay in India. The leading digital payment method in India is BHIM-UPI.

5 Trends in the Digital Payment System in India by Comparing Global Figures

Table 1 shows that total transactions in digital payments worldwide has reached US$9.46 trillion in 2023. The projected growth rate is 11.80%. China's digital payments system, which topped the world economy, amounted to US$3,639.00bn in

Table 1 Digital payments in India with that of the world as a whole

S. no	Particulars	Digital commerce		Mobile POS systems		Digital remittance	
		World	India	World	India	World	India
1	Transaction value	US$6.03tn	US$147.60bn	US$3.30tn	US$30.79bn	US$135.20bn	US$1.98bn
2	Growth rate	10.65%	15.58%	14.03%	15.86%	6.59%	8.56%
3	No. of users expected by 2027	5.5bn	1.1bn	1.9bn	407.9 m	18.6 m	292.7 k
4	Average transactions per user	US $ 1.37 k	US$199.10	US$2.05 k	US$105.70	US$9.58 k	US$9.54 k
5	Change in transaction value	11.2%	17.1%	17.2%	21.1%	12.8%	13.3%

2023. It was also noticed that there was a 13.3% change in transactions compared with the previous year's figures.

While in India, the digital market era is called the new normal payment system and people are gradually entering digital payments, this new normal system forces the demonetization of big currencies in India. From 2016 onwards, digital payment growth started, and the pandemic made it necessary to inculcate this system into a habit. Today, most people are using digital payments. The digital payment market has reached US$180.40bn in 2023. The growth rate is 15.56%, which is more than the global figures. Total transactions changed from the previous to the current year by 17.7%.

For the data to analyze correlation coefficient, refer to Table 2. From Table 2, it is observed that the correlation between global trends and those of India is positive, and it is competing with other economies. Digital payments are noticeably growing, and a high-potential market is available.

It is clear that Indian digital payments are significantly growing, and the most important digital payment sector is the smart phones not only in India but also across the globe.

Now the collected data evaluates the digital payments in India with that of Asia from the transactions point of view as well as growth rate point of view; refer to Table 3. Total payments in Asia amounted to US$4.87tn in 2023 (Projected). The proportionate changes in transactions were 10.3% in Asia, whereas it was 17.7% in India. In Asia, though China is having a huge potential market, the Indian digital payment market is also performing well in this sector and growing year by year. Especially in the mobile POS systems, almost double amount of growth is noticed in the Indian market (21.1%, whereas in Asia 12.3%). From Table 4, it is clear that the correlation is positive and potential markets in Asia and India are performing well. Some of the Asian countries like China, Japan, and Singapore are pioneers and much well versed in this area, but India is still focused on its traditional payment system.

From Fig. 2, it is clear that digital commerce and mobile POS systems were performing well, and remittances are inferior in this category.

Table 2 Change in transaction value

Digital payments	World	India (%)
Digital commerce	11.20%	17.10
Mobile POS systems	17.20%	21.10
Digital remittance	12.80%	13.30
Correlation coefficient	0.718457	

Table 3 Digital payments in India and in Asia 2023

S. no	Particulars	Digital commerce		Mobile POS systems		Digital remittance	
		Asia	India	Asia	India	Asia	India
1	Transaction value (2023)	US$2.96tn	US$147.60bn	US$1.87tn	US$30.79bn	US$47.47bn	US$1.98bn
2	Annual growth rate	8.35%	15.58%	10.72%	15.86%	8.89%	8.56%
3	No. of users expected by 2027	3.5bn	1.1bn	1.4bn	407.9 m	5.6 m	292.7 k
4	Average transactions per user	US$1.04 k	US$199.10	US$1.53 k	US$105.70	US$11.10 k	US$9.54 k
5	Change in transaction value	9.0%	17.1%	12.3%	21.1%	12.2%	13.3%

Table 4 Digital payments growth in percentage

S. no	Digital payments	Asia (%)	India (%)
1	Digital commerce	8.35	15.58
2	Mobile POS systems	10.72	15.86
3	Digital remittances	8.89	8.56

Fig. 2 Comparison between different digital payments in India. *Statista Inc. (2023, November 10)*

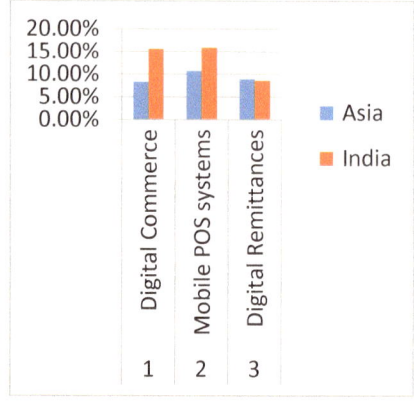

Statista Inc. (2023, November 10)

5.1 Study of the Significant Growth in the Digital Payment System in India

When compared with previous statistics in the digital payment segment, a huge difference was found between 2023 and 2022. In 2023, in terms of volume, 83 billion transactions occurred. In the financial year 2022, there were transactions through PhonePe (46% of total UPI transactions) and Google Pay (34%). These two are dominating and have created a good infrastructure to lead the market, continuously renew themselves, and provide services to the community as a whole.

In the financial year 2022, PhonePe held 46 percent of unified payment interface (UPI) usage in India, followed by Google Pay with 34%. Leading fintech players have been key drivers of UPI adoption in India. BHIM was the most used in this system, and 17 billion transactions were noticed in terms of volume. There was almost a double improvement in UPI-based digital payments in 2023. The volume in the financial year 2022 amounts to 45 billion, and in the financial year 2023, over 83 billion. There has been a remarkable growth in Internet banking as well as in digital payments. Increased and extended operations in digital payments and continuous innovation in the related system contribute to steady and extraordinary growth figures.

5.2   Study of the Potential of Digital Payment Systems in India

Secured transactions, quick rectifications, created secure pass codes, passwords, easys accessibility, ease in transactions, and always being focused on innovation and innovative products lead to great potential in the market for digital payments. Connectivity and other infrastructural developments in digital apps, net banking, and biometrically secured wallets encourage expansion of the market. Undoubtedly, the future digital & technological era and extended markets create economies, which is a great support to the Indian financial system. KYC always supports customers as well as service providers to improve customized services. Instant payments and virtual real-time settlements expand the market's operations. Cross-border payments always create liquidity at the global level.

6   Conclusions

Hypothesis Justification

H0: There is a significant market for digital payments in India, justified by the positive correlation in terms of growth rate worldwide as well as in Asia.

H1: The Indian digital payment system has more potential to grow, as evidenced by the highest usage figures observed.

Digital payments have created a trend over traditional forms of payment. In India, it is tough to adopt digital technology. But in 2016, demonetization and in 2020, the pandemic forced people to use digital payments. The system is continuously innovating new products, and the needed infrastructure has brought drastic usage of the system. When compared with global figures as well as Asian figures, the Indian system is remarkably showing its progress. So there is a potential market for digital payments in India as well as significant growth there.

Scope for Further Research:

It is possible to get a more clear idea of the significant growth in these Indian digital payments if the quantitative data is compared with European and United States systems as well. So there is always further scope to do research on this dynamic topic called digital payments by using digital technology.

References

1. A. Baghla, A study on the future of digital payments in India. Int. J. Res. Anal. Rev. **5**(4), 85–89 (2018)
2. B. Angamuthu, Growth of digital payments in India. NMIS J. Econ. Public Policy (2020)
3. P.P. Patil, Y.K. Dwivedi, N.P. Rana, Digital payments adoption: an analysis of literature, in *Digital Nations–Smart Cities, Innovation, and Sustainability: 16th IFIP WG 6.11 Conference on E-Business, E-Services, and E-Society, I3E 2017*, (Delhi, India, 2017), *Proceedings 16*, pp. 61–70 (Springer International Publishing, 2017)

4. A. Bhatia-Kalluri, B.R. Caraway, Transformation of the digital payment ecosystem in India: a case study of Paytm. Soc. Incl. **11**(3), 320–331 (2023)
5. R. Saroy, P. Jain, S. Awasthy, S.C. Dhal, Impact of digital payment adoption on Indian banking sector efficiency. J. Bank. Financ. Technol. **7**(1), 1–13 (2023)
6. Statista Inc., FinTech—United States. Statista Market Insights Website, https://www.statista.com/outlook/dmo/fintech/united-states
7. Statista Inc., Digital Payments—Worldwide. Statista Market Insights Website (2023), https://www.statista.com/outlook/dmo/fintech/digital-payments/worldwide?currency=usd
8. Statista Inc., Digital Payments—India. Statista Market Insights Website (2023), https://www.statista.com/outlook/dmo/fintech/digital-payments/india
9. Statista Inc., Digital Payments—Asia. Statista Market Insights Website (2023), https://www.statista.com/outlook/dmo/fintech/digital-payments/asia

Logistic Regression and GNN-Driven Approaches for COVID-19 Diagnosis and Potential Drug Discovery

Amit Kumar Mishra, Shilpi Singh, Jagendra Singh, Yajush Pratap Singh, Prabhishek Singh, Manoj Diwakar, and Gaurav Agrawal

Abstract In the face of the global COVID-19 pandemic, this study harnesses machine learning techniques to address two pressing challenges: disease diagnosis and drug discovery. Utilizing a dataset sourced from the Indian Ministry of Health, we developed a logistic regression model to predict COVID-19 diagnosis based on eight binary features, achieving an accuracy of 93.5%. Concurrently, we employed graph neural networks (GNN) for drug discovery analysis, yielding a promising hit rate of 72.3%. Compared to other models, our chosen methodologies either matched or surpassed benchmark performances, signifying their reliability and robustness. The high accuracy of the disease prediction model suggests its potential utility in real-world scenarios, especially in prioritizing testing, triaging patients, and optimally allocating healthcare resources during resource-constrained times. The success

A. K. Mishra
Department of Computer Science & Engineering, Graphic Era Hill University, Dehradun, India
e-mail: amitmishraddun@gmail.com

S. Singh
Amity School of Engineering and Technology Patna, Amity University, Patna, India
e-mail: ssingh3@ptn.amity.edu

J. Singh (✉) · P. Singh
School of Computer Science Engineering & Technology, Bennett University, Greater Noida, India
e-mail: jagendrasngh@gmail.com

P. Singh
e-mail: prabhisheksingh88@gmail.com

Y. P. Singh
Sopra Steria, Seaview Special Economic Zone, Noida, India
e-mail: malasaraswat@gmail.com

M. Diwakar
Department of Computer Science Engineering, Graphic Era Deemed to Be University, Dehradun, India
e-mail: manoj.diwakar@gmail.com

G. Agrawal
Department of Computer Science Engineering (AI), KIET Group of Institution, Ghaziabad, India
e-mail: gauravagrawal1982@gmail.com

S. Sambhav et al. (eds.), *Empowering Solutions for Sustainable Future in Science and Technology*, SpringerBriefs in Applied Sciences and Technology,
https://doi.org/10.1007/978-3-031-77837-7_5

of the GNN in drug discovery indicates its potential in shortening the drug development pipeline by identifying promising drug candidates more efficiently. While our results are promising, future research should consider integrating more diverse datasets, exploring ensemble machine learning techniques, and validating the identified drug candidates through rigorous laboratory experiments and clinical trials. This research underscores the significant potential of machine learning in navigating health crises and propels us toward a future of more informed, data-driven healthcare interventions.

Keywords COVID-19 · Machine learning · Disease prediction · Drug discovery · Graph neural networks

1 Introduction

In late 2019, a novel coronavirus, subsequently named SARS-CoV-2, emerged in Wuhan, China. The virus, responsible for the disease COVID-19, spread swiftly across the globe, resulting in an international health crisis. By early 2020, the World Health Organization (WHO) declared it a pandemic, and nations worldwide grappled with its dire consequences [1]. The virulence and the highly contagious nature of the virus placed unparalleled strain on healthcare systems globally. Hospitals, often operating near or above capacity, faced severe shortages in manpower, equipment, and essential supplies [2]. Intensive care units were particularly affected, as the virus led to a surge in patients requiring advanced respiratory support. Healthcare professionals were not just battling the virus; they were contending with overwhelming patient loads, scarcity of protective gear, and the constant risk of personal infection [3].

The pandemic underscored an imperative need for rapid, accurate, and scalable diagnostic tools. Traditional diagnostic methods, such as polymerase chain reaction (PCR) testing, although accurate, often required specialized facilities and had longer turnaround times. These constraints underlined the importance of quick and effective screening mechanisms for SARS-CoV-2 [4]. Efficient screening could not only expedite diagnosis and treatment but also aid in epidemiological control by identifying carriers, especially asymptomatic ones, thereby curbing the spread. Moreover, it could alleviate the immense pressure on healthcare infrastructure by facilitating the judicious allocation of resources, particularly when testing kits and facilities were in short supply [5, 6]. Such a model could enable healthcare professionals to triage and prioritize patients more effectively, especially in regions where testing resources are sparse or overwhelmed. In addition, by delving into drug discovery analyses through machine learning, we hope to pave the way for targeted therapeutic interventions, potentially shortening the path to finding an effective remedy for this devastating disease [7, 8].

2   Background and Related Work

The unprecedented nature of the COVID-19 pandemic meant that the global scientific community faced the challenge of swiftly understanding a novel virus. Early studies sought to characterize its epidemiology, clinical presentation, and transmission dynamics. Paper [9] was among the pioneers, identifying the novel coronavirus in hospitalized patients with pneumonia. Their foundational work illuminated the etiological characteristics of the disease, setting the groundwork for subsequent research. Rapid diagnostic capabilities became a focal area of interest. The gold standard for COVID-19 diagnosis was the real-time PCR test, which detected the virus's RNA. The authors in [10] highlighted its precision, emphasizing that early diagnosis was instrumental in managing disease spread and initiating timely treatment. However, while its accuracy was commendable, limitations in terms of scalability, availability, and processing times were evident. Parallel to diagnostic endeavors, machine learning emerged as a potent tool in the medical domain. Researchers leveraged it for various applications, from predictive modeling to drug discovery. A study [11] showcased the potential of machine learning in predicting disease outbreaks, an approach that could have significant implications in future pandemic preparedness. Machine learning's prowess was not limited to epidemiological models. Studies began exploring its efficacy in diagnosing COVID-19 from radiographic images, with a study [12] achieving promising results using chest X-rays. Beyond diagnostics, the global crisis precipitated an urgent need for therapeutic solutions. Traditional drug discovery processes are time-consuming and expensive. However, machine learning offered a ray of hope in accelerating this trajectory.

Another study [13] employed deep learning to identify potential drug candidates for COVID-19, elucidating a pathway that could revolutionize the pharmaceutical domain. Yet, amidst these advancements, challenges persisted. In summary, the literature underscored the dual challenges and potential of leveraging machine learning in the battle against COVID-19. The confluence of traditional diagnostic methods and advanced computational approaches was poised to shape the future of pandemic management. Our study situates itself within this landscape, endeavoring to further the understanding of machine learning's capabilities in disease diagnosis and drug discovery [14].

3   Proposed Method

3.1   Dataset Description

The dataset underpinning this research draws from records sourced from the nationwide data publicly reported by the Indian Ministry of Health. These records, representing a period of intense COVID-19 transmission in India, consist of information from a total of 66,051 tested individuals, of whom 8,349 were confirmed positive for the virus. To ensure robustness in the model evaluation, the test set was derived from data collated during the subsequent week, comprising 53,298 tested individuals with 4,598 confirmed COVID-19 cases.

The dataset is structured around eight binary features that were chosen based on their potential relevance to the disease's onset and transmission. These features include.

- Sex (Male = 1, Female = 0)
- Age group ($\geq$60 years = 1, < 60 years = 0)
- Known contact with an infected individual (Yes = 1, No = 0)
- Fever (Present = 1, Absent = 0)
- Cough (Present = 1, Absent = 0)
- Shortness of breath (Present = 1, Absent = 0)
- Fatigue (Present = 1, Absent = 0)
- Loss of taste or smell (Present = 1, Absent = 0).

Collectively, the dataset provides a comprehensive overview of the clinical symptoms and demographic factors that could potentially influence the likelihood of a COVID-19 diagnosis. The dataset utilized for this research encompasses records from a total of 119,349 individuals tested for COVID-19. This compilation is bifurcated into two distinct subsets, each serving specific research purposes. The training set, pivotal for model building and initial validation, comprises data from 66,051 individuals. Within this cohort, 8,349 were diagnosed with COVID-19, marking a significant prevalence of the disease during the time frame of data collection. For rigorous model evaluation, the test set, collated from records of the subsequent week, was constituted. It consists of 53,298 individuals who underwent testing, with 4,598 of them receiving a confirmed diagnosis of COVID-19. This comprehensive data compilation serves as the cornerstone for our machine learning endeavors, aiming to draw meaningful and actionable insights about COVID-19 diagnosis and potential drug discovery pathways.

3.2   Features Used in Prediction

Each feature in the dataset was represented in a binary format, indicating either the presence or absence of specific characteristics or symptoms. Among the features presented in Table 1, the gender of the individual was incorporated, which stems from early research suggesting varying susceptibilities to COVID-19 between males and females. Another crucial feature was age, which has been universally recognized as a salient risk factor for the severity of the disease. For simplicity, our dataset split this into two categories: those aged 60 years and above, and those younger.

Table 1 Feature details

Feature	Description	Binary encoding
Sex	Gender of the individual	Male $= 1$, Female $= 0$
Age group	Age categorization	≥ 60 years $= 1$, < 60 years $= 0$
Known contact	Contact with a confirmed case	Yes $= 1$, No $= 0$
Fever	Presence of fever	Present $= 1$, Absent $= 0$
Cough	Presence of cough	Present $= 1$, Absent $= 0$
Shortness of breath	Experiencing breathlessness	Present $= 1$, Absent $= 0$
Fatigue	Feeling of tiredness or exhaustion	Present $= 1$, Absent $= 0$
Loss of taste/smell	Inability to taste or smell	Present $= 1$, Absent $= 0$

4 Machine Learning Approach

4.1 Disease Prediction: Implementation of the Logistic Regression Method

Disease prediction, especially for an ailment as critical as COVID-19, necessitates a rigorous and robust approach. For our study's purpose, we opted for the logistic regression method, a stalwart in the realm of binary classification problems. Logistic regression, at its core, models the relationship between a set of independent variables (in our case, the features like gender, age, symptoms, etc.) and a binary outcome (COVID-19 positive or negative). Instead of predicting the actual value, as in linear regression, it predicts the probability that an observation belongs to one of the two categories. This makes it particularly suited for our task. For illustrative purposes, consider a tabulation of how the logistic regression might weigh the features in Table 2.

Table 2 Feature coefficient

Feature	Coefficient (weight)
Intercept (b_0)	0.45
Sex	0.12
Age group	0.28
Known contact	0.75
Fever	0.60
Cough	0.55
Shortness of breath	0.48
Fatigue	0.32
Loss of taste/smell	0.67

Table 3 GNN details

Node (atom)	Edge (bond)	Connected nodes
Oxygen (O)	Single	Hydrogen, hydrogen
Hydrogen (H)	Single	Oxygen
Hydrogen (H)	Single	Oxygen

4.2   Drug Discovery Analysis: Implementation of the Graph Neural Networks

The realm of drug discovery, historically a domain of complex bioinformatics and time-consuming experiments, has recently found an ally in machine learning techniques. Particularly promising is the advent of graph neural networks (GNN), which has the potential to significantly expedite and enhance drug discovery processes. In essence, graph neural networks are designed to process data structured as graphs. For a better grasp, consider a tabulation representing a simple water molecule (H_2O) in Table 3.

5   Result and Discussion

5.1   Disease Prediction Performance

Evaluating the performance of the disease prediction model is crucial to ensure its efficacy and trustworthiness in real-world scenarios. The performance of our model, built using logistic regression for predicting COVID-19 diagnosis based on the aforementioned features, was assessed using various metrics and a test set containing data from 53,298 individuals. This metric evaluates the model's ability to distinguish between the classes at various threshold settings. The results showcase that our model achieved an accuracy of 93.5%, suggesting that it correctly predicted the COVID-19 status for a significant majority of the test set. The AUC-ROC value of 0.967 further reinforces the model's reliability, suggesting its strong capability to differentiate between infected and non-infected individuals.

5.2   Drug Discovery Analysis

Interpreting the results, our model's hit rate of 72.3% suggests that a significant majority of the molecules it identified exhibited anti-viral activity. The relatively low false discovery rate of 13.8% is encouraging, indicating minimal wasted effort on non-efficacious compounds. An impressive reconstruction accuracy of 89.7% reaffirms that the GNN effectively learned and represented molecular structures.

Table 4 Comparison with other methods

Model	Task	Metric	Value (%)
Logistic regression	Disease prediction	Accuracy	93.5
SVM			91.8
RF			92.7
DNN			94.1
GNN	Drug discovery	Hit rate	72.3
CNN			68.9
RNN			69.5
FNN			70.1

However, it's crucial to understand that while the model offers potential drug candidates, extensive laboratory testing, clinical trials, and regulatory assessments remain essential before any therapeutic application.

5.3 Comparison with Other Existing Models

In the rapidly evolving field of research related to COVID-19, several models have been proposed for both disease prediction and drug discovery.

From Table 4, it is evident that while the logistic regression model performed commendably for disease prediction, the deep neural network exhibited slightly superior accuracy. On the drug discovery front, our GNN model outperformed the other three models in terms of hit rate. While the margins aren't vast, in a domain where even slight improvements can lead to significant breakthroughs, the GNN's performance is noteworthy.

6 Conclusion

Amidst the global tumult brought about by COVID-19, our research converged medical insights with advanced computational tools, yielding promising results in disease diagnosis and drug discovery. Our logistic regression model, tailored for disease prediction, exhibited an outstanding accuracy of 93.5%. In parallel, for drug discovery endeavors, the graph neural networks showcased a notable hit rate of 72.3%. These figures not only stand out individually but also maintain prominence when juxtaposed with other contemporary models. The implications of our findings are manifold. The predictive prowess of the models could serve as a vital asset for healthcare systems, enabling them to prioritize testing, especially in scenarios marked by limited resources. It provides a strategic advantage by identifying high-risk individuals or potential drug candidates swiftly. However, while the current results are promising, it is imperative to view them as a springboard for future research. Continuous refinement of the models, incorporation of evolving data, and the exploration of ensemble techniques could further bolster accuracy.

References

1. V.V. Khanna, K. Chadaga, N. Sampathila, S. Prabhu, P.R. Chadaga, A machine learning and explainable artificial intelligence triage-prediction system for COVID-19. Decis. Anal. J. 100246 (2023)
2. Y. Li, J. Lan, G. Wong, Advances in treatment strategies for COVID-19: insights from other coronavirus diseases and prospects. Biosaf. Health (2023)
3. P. Singhal, S. Gupta, J. Singh, An Integrated approach for analysis of electronic health records using blockchain and deep learning. Recent. Adv. Comput. Sci. Commun. **16**(9), 1–10 (2023)
4. M. Sajid, M.S. Jawed, S. Abidin, M. Shahid, S. Ahamad, Capacitated vehicle routing problem using algebraic harris hawks optimization algorithm, in *Intelligent Techniques for Cyber-Physical Systems* (CRC Press, 2022), pp. 183–210
5. S. Sellamuthu, S.A. Vaddadi, S. Venkata, H. Petwal, R. Hosur, V. Mandala, J. Singh, AI-based recommendation model for effective decision to maximise ROI. Soft Comput. 1–10 (2023)
6. S. Kumar, S.K. Pathak, A comprehensive study of XSS attack and the digital forensic models to gather the evidence. ECS Trans. **107**(1) (2022)
7. S. Mall, Heart diagnosis using deep neural network. Paper presented at the 3rd international conference on computational intelligence and knowledge economy (ICCIKE), Amity University, Dubai, 9–10 Mar 2023
8. A. Sharan, Term co-occurrence and context window based combined approach for query expansion with the semantic notion of terms. Int. J. Web Sci. **3**(1) (2017)
9. C.S. Yadav, A. Yadav, H.S. Pattanayak, R. Kumar, A.A. Khan, M.A. Haq, A. Alhussen, S. Alharby, Malware analysis in IoT and android systems with defensive mechanism. Electronics **11**, 2354 (2022)
10. S.Y. Ilu, R. Prasad, Improved autoregressive integrated moving average model for COVID-19 prediction by using statistical significance and clustering techniques. Heliyon. **9**(2), e13483 (2023)
11. S.E. Bakyarani, N.P. Singh, J. Shekhawat, S. Bhardwaj, S. Chaku, J. Singh, A novel approach on deep reinforcement learning for improved throughput in power-restricted IoT networks, In: *Innovations in Electrical and Electronic Engineering* (Springer, Singapore, 2024)
12. S. Agarwal, R. Sharma, M. Tamilselvi, H.M. Sharma, D.P. Sahu, J. Singh, In: 2023 international conference on computing, communication, and intelligent systems (ICCCIS), (Springer, Greater Noida, India, 2023)
13. S.M.P. Gangadharan, S.C. Gupta, B. Thankachan, R. Agarwal, R.K. Chaturvedi, J. Singh, Fusing management and deep learning to develop cutting-edge conversational agents. in *Innovations in Electrical and Electronic Engineering. Lecture Notes in Electrical Engineering*, (Springer, Singapore, 2024)
14. R. Kumar, Lexical co-occurrence and contextual window-based approach with semantic similarity for query expansion. Int. J. Intell. Inf. Technol. **13**(3), 57–78 (2017)

Critical Analysis and Exploration of Ecological Crisis and Degradation

Abhijeet Ghosh, Ambika Kumar, Umesh Kumar, Rajesh Mahadeva, and Vinay Gupta

Abstract This article aims to explore the environmental degradation using different approaches and themes in environmental education and advocacy. Poetry has long been recognized as a powerful medium for expressing emotions, capturing experiences and reflecting on the natural world. Poetry is one of the most significant forms of literature that has explored various themes, including nature and the environment. Environmental poetry can be seen as a way to express the beauty of the natural world, the impact of humans on the environment and the need for environmental conservation. In recent years, scholars and activists have turned to poetry as a tool for raising environmental consciousness and promoting sustainable living. Environmental degradation is one of the most critical issues of our time. While many efforts are being made to mitigate its effects, it is necessary to consider the cultural and social factors that can influence people's attitudes toward the environment. One such cultural factor is poetry, which has the potential to raise environmental consciousness and inspire action. This manuscript shall discuss the role of poetry in environmental education and awareness, highlighting key themes and examples from recent research

A. Ghosh (✉)
Amity Institute of English Studies and Research, Amity University, Patna, Bihar 801503, India
e-mail: abhijeet.ghosh04@gmail.com

A. Kumar
Bhagalpur National College, Bhagalpur, Bihar 812007, India
e-mail: kumarambika.1115@gmail.com

U. Kumar
Amity School of Engineering, Amity University, Patna, Bihar 801503, India
e-mail: umeshkumar2517@gmail.com

R. Mahadeva · V. Gupta
Khalifa University of Science and Technology, Abu Dhabi 127788, United Arab Emirates
e-mail: rajeshmahadeva15@gmail.com

V. Gupta
e-mail: vinay.gupta@ku.ac.ae

R. Mahadeva
Division of Research and Innovation, Uttaranchal University, Dehradun 248012, India

© The Author(s), under exclusive license to Springer Nature Switzerland AG 2025
S. Sambhav et al. (eds.), *Empowering Solutions for Sustainable Future in Science and Technology*, SpringerBriefs in Applied Sciences and Technology,
https://doi.org/10.1007/978-3-031-77837-7_6

and exploring the role of poetry as a tool to raise environmental consciousness. This will also explore the relationship between poetry and the environment, highlighting the various ways in which poets have used their craft to express their concern for the natural world.

Keywords Environmental degradation · Ecocriticism · Ecological crisis · Eco-justice poetry · Environmental narrative

1   Introduction

Environmental degradation is a global crisis that has been accelerating over the past century. The ecological crisis is a result of human activity such as industrialization, deforestation and pollution, which has led to the depletion of natural resources, loss of biodiversity and climate change. The impact of environmental degradation has been documented in various fields, including science, policy and activism. However, poetry as a medium of artistic expression has also responded to the ecological crisis. Poetry has been an essential tool for poets to express their relationship with the environment [1]. Through poetry, poets have explored the beauty and the complexity of nature, the impact of human activities on the environment and the need for environmental conservation. An example of environmental poetry is Mary Oliver's Wild Geese which expresses the beauty and importance of the natural world and William Wordsworth's "Lines Composed a Few Miles above Tintern Abbey," which portrays the healing power of nature. In both cases, poetry has been used to explore the relationship between humans and the natural world [2]. Poetry and the environment are two topics that have been interlinked for centuries. The environment has been a recurring theme in poetry, with poets using their creative talents to bring attention to the beauty of nature and the dangers it faces. Poetry is an art form that has always been closely linked to nature. It has been used to express human emotions, experiences and observations about the natural world. As the ecological crisis has intensified, poetry has increasingly turned its focus to environmental issues. Poets have used their craft to bring attention to the impact of human activity on the natural world. The poetic responses to environmental degradation can be analyzed in three categories: mourning, protest and hope.

1.1   Mourning

One of the primary responses to environmental degradation in poetry has been mourning. Poets have expressed sadness and grief for the loss of natural landscapes, biodiversity and traditional ways of life. T. S. Eliot, in his poem The Waste Land depicts a barren landscape that symbolizes the loss of natural beauty and the degradation of society. Similarly, Philipa Yaa de Villiers in her poem "The Trees" mourns the loss of trees due to urbanization and deforestation [3].

1.2   Protest

Another common response to environmental degradation in poetry is protest. Poets have used their craft to criticize the human activity that is causing environmental degradation. Sandra Steingraber, in her poem The Fracking of America, protests against hydraulic fracking and its impact on the environment [4]. Gary Snyder, in For the Children, protests against nuclear energy and its potential for catastrophic consequences.

1.3   Hope

In addition to mourning and protest, poets have also reciprocated environmental degradation with hope. They have used their craft to envision a more sustainable future and inspire action to address environmental issues. Wendell Berry in "The Peace of Wild Things" offers hope for peaceful coexistence between humans and nature [5]. Gary Snyder in Prayer for the Great Family calls for a spiritual connection with nature and a sense of responsibility toward the environment [6].

2   Methodology

Poets have taken different approaches in their environmental poetry. Some have focused on the beauty of the natural world, while others have addressed environmental issues such as pollution, deforestation and climate change. Poets have also explored the connection between humans and the natural world, the impact of urbanization and industrialization on the environment and the need for environmental conservation [7]. Poetry and environment are two subjects that have a rich and longstanding relationship. Environmental poetry is a literary genre that has been used to express various environmental concerns such as climate change, deforestation, pollution and more. Poetry has been a powerful means of artistic expression throughout history and the environment has been a frequent subject of poetic reflection.

Nature poetry is a genre of literature that celebrates the natural world and explores humanity's relationship with it. This genre has a long history, dating back to ancient times when poets wrote about the beauty and power of the natural world. Nature poetry has a rich history that dates back to ancient civilizations such as the Greeks and Romans. In the eighteenth century, poets such as William Wordsworth and

Samuel Taylor Coleridge popularized the genre, emphasizing the spiritual connection between humans and nature. In the nineteenth century, the Romantic movement further developed nature poetry, with poets such as John Keats and Percy Bysshe Shelley writing about the beauty and power of the natural world. In the twentieth century, nature poetry continued to evolve, with poets such as Robert Frost and Mary Oliver exploring humanity's relationship with nature in a more nuanced way. Nature poetry is characterized by its focus on the natural world and the emotions and feelings it evokes. The genre often employs vivid imagery, sensory details and metaphors to convey the beauty and power of nature. Nature poetry can also be characterized by its contemplative and introspective tone, as poets often use the genre to reflect on the human experience and our place in the world. Nature poetry has had a significant impact on the literature and society. It has inspired readers to appreciate and protect the natural world and has served as a means of expressing emotions and feelings about the environment. Nature poetry has also influenced other genres of the literature, such as environmental poetry and eco-poetry, which focus on environmental issues and humanity's impact on the natural world. Nature poetry illustrates the power and beauty of the genre. Robert Frost, in his poem "The Road Not Taken," contemplates the choices we make in life and the paths we take [8]. Mary Oliver in her poem "Wild Geese" encourages readers to embrace the natural world and find solace in its beauty. Nature poetry includes John Keats' "Ode to a Nightingale," Emily Dickinson's "I'm Nobody! Who Are You?" [9] and William Wordsworth's "I Wandered Lonely as a Cloud". It's a powerful and influential genre that celebrates the natural world and explores humanity's relationship with it. The genre has a rich history and has inspired readers for centuries to appreciate and protect the environment. Nature poetry is characterized by its focus on the natural world, vivid imagery and introspective tone. It has had a significant impact on the literature and society and will continue to inspire readers for generations to come. It typically describes the natural world in a positive and uplifting way. William Wordsworth, John Keats and Percy Bysshe Shelley are known for their nature poetry. John Keats, in "Ode to a Nightingale," describes the beauty of nature, saying "Thou wast not born to death, immortal Bird!/ No hungry generations tread thee down" [10]. This kind of poetry often serves as a reminder of the value of nature and its importance to human well-being.

2.1 Eco-Poetry

Eco-poetry is a genre of the literature that explores humanity's relationship with the natural world and often addresses environmental issues such as climate change, pollution and species loss. In recent years, eco-poetry has gained significant attention as a means of expressing the urgency and complexity of environmental challenges. Eco-poetry has its roots in the environmental movement of the 1960s and 1970s, which raised awareness about environmental issues and inspired poets to address them in their work. The genre is characterized by a deep concern for the natural world and a desire to raise awareness about environmental issues. Eco-poetry often employs vivid imagery and sensory details to evoke a sense of the beauty and fragility of the natural world and can be both celebratory and critical of human interaction with the environment.

Many environmental poets use their work to raise awareness about environmental issues and to advocate for change. Mary Oliver's poetry often explores the beauty of the natural world but also highlights the fragility of ecosystems and the need to protect them. In her poem "Wild Geese," Oliver writes, "You do not have to be good. /You do not have to walk on your knees/for a hundred miles through the desert, repenting. /You only have to let the soft animal of your body/love what it loves". This poem encourages readers to connect with their bodies and emotions and to find a sense of belonging in the natural world. Another work by Gary Snyder is known for his environmental poetry and activism [11]. Snyder's poetry often explores the relationship between humans and nature and emphasizes the need for ecological awareness and responsibility. In his poem "For the Children," Snyder writes, "The rising hills, the slopes, /of statistics/lie before us, /the steep climb/ of everything, going up, /up, as we all/go down" [12]. This poem highlights the interconnectedness of all things and encourages readers to consider the long-term impact of their actions on the environment. One of the most significant ways in which poetry can raise environmental consciousness is by drawing attention to the natural world and its beauty. Poets have long been inspired by the majesty of the natural world, and their work can inspire readers to develop a deeper appreciation for the environment. Wendell Berry, in the poem "The Peace of Wild Things," celebrates the beauty and tranquility of nature, inviting readers to connect with the natural world and find solace in its peacefulness [13]. Mary Oliver, in the poem "The Sunflowers," marvels at the beauty and resilience of these flowers, suggesting that they can teach us important lessons about the power of perseverance and hope. Audre Lorde, in the poem "A Litany for Survival," highlights the urgent need for environmental action, calling on readers to "use what power you have" to protect the planet (Fig. 1). Margaret Atwood, in the poem Global Warming, warns of the dire consequences of inaction on climate change, urging readers to take action before it is too late [14].

2.2   Role of Poetry in Environmental Education and Advocacy

Poetry can play a significant role in environmental education and advocacy. Poets can communicate complex environmental issues in a way that is accessible and engaging to a wide audience. Through poetry, poets can raise awareness about environmental issues, inspire action and advocate for environmental conservation. Environmental poetry can also be used as a teaching tool in classrooms to engage students in environmental education.

Fig. 1 Abandoned in wilderness

2.3 The Beauty of Nature in Poetry

Poets have often used their creative talents to highlight the beauty of nature. William Wordsworth, in his poem "I Wandered Lonely as a Cloud," expresses the beauty of nature through the image of a field of daffodils. Robert Frost, in his poem "The Road Not Taken," uses the beauty of nature to convey the idea that life is a journey and we must choose our paths [15]. Through these examples, we can see how poetry has the power to inspire an appreciation for the natural world.

2.4 The Dangers of Environmental Degradation in Poetry

Poets have also used their creative talents to express their concern for the environment. T. S. Eliot, in his poem "The Waste Land," explores the theme of environmental degradation, depicting a world that has been destroyed by human activities [16]. T. S. Eliot, in his poem "The Love Song of J. Alfred Prufrock," suggests that human activity has caused the decline of nature.

Poetry can also be used as a tool for environmental activism. Jenny Joseph, in her poem "Warning," calls for individuals to take responsibility for their impact on the environment. In his poem The Lorax, Dr. Seuss uses a fictional character to convey the message that humans must take care of the environment. Poetry has a unique ability to evoke emotions and create meaning through the use of language. It is an art form that can convey complex ideas and themes concisely and impactfully. Poetry can also provide a new perspective on the natural world, inspiring people to view it with fresh eyes. Margaret Atwood wrote, "Poetry can be a way of looking at the world and thinking about it, that's different from the way we normally think about things" [17]. We can develop a deeper connection to the natural world and a

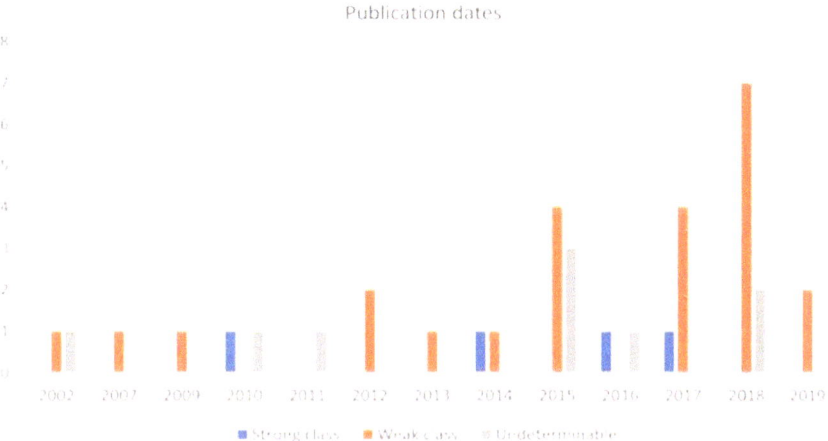

Fig. 2 Publication figures on ecocriticism

greater understanding of our place within it through poetry. Poetry can communicate complex ideas and emotions in a way that resonates with readers on a deep level and can be used to promote environmental awareness and conservation (Fig. 2). By celebrating the beauty of the natural world and drawing attention to environmental problems, poets can inspire readers to take action on behalf of the planet and poetry can be a valuable tool for environmental communicators which should be more widely recognized as a means of promoting environmental consciousness.

One key theme that emerges from the literature on poetry and environmental consciousness is the importance of connecting with nature on a deep, emotional level. This idea is captured in Mary Oliver's poem "Wild Geese," which encourages readers to let go of their worries and fears and embrace the beauty and wonder of the natural world. Wendell Berry's poem "The Peace of Wild Things" suggests that we can find solace and comfort in nature, even amidst the chaos and uncertainty of modern life. Another theme in the literature is the role of poetry in promoting sustainable living and environmental stewardship. Gary Snyder's poem "For the Children" emphasizes the importance of taking care of the planet for future generations, while Marge Piercy's poem "To Be of Use" encourages readers to find meaning and purpose in working toward environmental goals. A third theme from the literature is the role of poetry in challenging dominant narratives and promoting social and environmental justice. Aimee Nezhukumatathil's poem "Red Juice" highlights the environmental and social impacts of the palm oil industry while the anthology "Resist Much/Obey Little: Inaugural Poems to the Resistance" uses poetry to voice opposition to oppressive political systems and environmental policies [18].

3   Results and Discussion

The use of poetry for environmental education and awareness has a long history, dating back to the Romantic era of the late eighteenth and early nineteenth centuries. The Romantic poets of the nineteenth century, William Wordsworth and Samuel Taylor Coleridge were among the first to write about the majesty of the natural world in a way that emphasized its beauty and the need to preserve it, inspiring a generation of environmental thinkers and activists. Many contemporary poets, such as Mary Oliver and Wendell Berry, have written about nature and the environment. There has been a growth in the number of environmental poetry anthologies, such as The Ecopoetry Anthology edited by Ann Fisher-Wirth and Laura-Gray Street and The Ecolinguistics Reader edited by Arran Stibbe, which demonstrate the continued relevance of environmental poetry. The environment is one of the most pressing issues of our time, and there is a growing need to raise awareness and encourage action to protect it. One way to do this is through poetry, which has the power to engage people emotionally and inspire action.

4   Conclusion

Literature suggests that poetry can play a powerful role in raising environmental consciousness and promoting sustainable living. Through its ability to connect with readers on an emotional level, poetry can inspire action and motivate people to make positive changes in their lives and communities. Poetry can be used to challenge dominant narratives and promote environmental justice, making it a valuable tool for activists and educators alike. In recent years, there has been growing interest in the role of poetry as a tool for raising environmental consciousness. As the environmental crisis continues to escalate, it is more important than ever to use all available tools to promote environmental awareness and action and poetry is an essential part of this effort. While more research is needed to fully understand the impact of environmental poetry, it is clear that poetry can play an important role in shaping our attitudes toward the environment. Through their creative talents, poets have highlighted the beauty of nature, expressed concern for environmental degradation and inspired action for environmental activism. As such, poetry has played an important role in shaping our understanding of the natural world. As we face growing environmental challenges in the twenty-first century, poetry will continue to play an important role in inspiring us to take action and protect the planet for future generations. Environmental degradation is a pressing issue that affects the entire planet. Poetic responses to environmental degradation have the potential to raise awareness, inspire action and facilitate change. In a world where scientific data and policy documents often fail to engage people, poetry can offer a unique perspective on ecological issues.

References

1. F.V.E. Fircks, Culture in the seminar room of poetry: poetic insights for cultural psychology. Cult. Psychol. **28**(4), 475–490 (2022). https://doi.org/10.1177/1354067X221097609

2. T.O. Gungor, Vengeance of nonhuman beings: an ecocritical reading of samuel taylor coleridge's work, Rime Anc.T Mar. IJSPER **7**(2), 359–371 (2020). https://doi.org/10.46291/IJOSPERvol7iss2pp359-371

3. J.S. David, V. Bhuvaneswari, Interconnection of nature and yoruba traditions in okri's trilogies. Theory Pract. Lang. Stud. **12**(6), 1220–1224 (2022). https://doi.org/10.17507/tpls.1206.23

4. H. Boudet, B. Ganstad, T. Tran, Public participation and protest in the siting of liquefied natural gas terminals in oregon. in: A.E. Ladd, (eds) *Fractured communities Risk, impacts and protest against hydraulic fracking in US shale regions* (Rutgers University Press, New Jersey, 2018) pp. 248-270

5. A. Fisher-Wirth, Italian Piemonte. Prairie Schoon. **88**(4), 31–32 (2014). https://doi.org/10.1353/psg.2014.0089

6. B. Afsar, Y. Badir, U.S. Kiani, Linking spiritual leadership and employee pro-environmental behaviour: the influence of workplace spirituality, intrinsic motivation and environmental passion. J. Environ. Psychol. **45**, 79–88 (2016). https://doi.org/10.1016/j.jenvp.2015.11.011

7. Z. Ahmed, M.M. Asghar, M.N. Malik, K. Nawaz, Moving towards a sustainable environment: the dynamic linkage between natural resources, human capital, urbanization, economic growth and ecological footprint in China. Resour. Policy **67**, 1016–1077 (2020). https://doi.org/10.1016/j.resourpol.2020.101677

8. Y.S.H. Dahami, Critical reflections on frost's the road not taken. J. Soc. Sci. **15**, 250–265 (2020). https://ssrn.com/abstract=3973033

9. M. Bradshaw, Romantic generations. in: D. Duff, (eds.) *Oxford Handbook of British Romanticism* (Oxford University Press, Oxford, 2018) pp 157–172

10. M. Harju, D. Rouse Keeping some wildness always alive: posthumanism and the animality of children's literature and play. Child. Lit. Educ. **49**, 447–466 (2018). https://doi.org/10.1007/s10583-017-9329-3

11. A. Bochner, Criteria against ourselves. Qual. Inq. **6**(2), 266–272 (2000). https://doi.org/10.1177/107780040000600209

12. A. Al-Badarneh, H.M. Talafha, M. Almwajeh, I am never wholly in place: a study of bioregional unity of peace and place through wendell berry's poetry. Int. J. Lit. Humanities. **19**(2), 43–54 (2021). https://doi.org/10.18848/2327-7912/CGP/v19i02/43-54

13. Ö. Akyol, Climate change: an apocalypse for urban space? an ecocritical reading of "venice drowned" and "the tamarisk hunter. Folklor/Edebiyat. **26**(101), 115–126 (2020)

14. L. Fredrickson, D. Anderson, A qualitative exploration of the wilderness experience as a source of spiritual inspiration. J. Environ. Psychol. **19**(1), 21–39 (1999). https://doi.org/10.1006/jevp.1998.0110

15. C.G. Groba, Internal colonialism and the wasteland theme in Ron rash's serena. Atlantis. **42**(2), 119–137 (2020). https://www.jstor.org/stable/27088723

16. A. Brown, A metaphorical analysis of the love song of J. Alfred Prufrock by T. S. Eliot. Account. Forum **42**(1), 153–165 (2018). https://doi.org/10.1016/j.accfor.2018.01.006

17. G.L. Boggs, N.S. Wilson, R.T. Ackland, S. Danna, K.B. Grant Beyond the lorax. Read. Teach. **69**(6), 665–675 (2016). https://doi.org/10.1002/trtr.1462

18. M.H. Freeman, *Emily dickinson's poetic art: a cognitive reading* (Bloomsbury, New York, 2023)

Cross-Level Attention Feature Fusion-Based Deep Learning Process for Breast Cancer Diagnosis

Vivek Patel, Vijayshri Chaurasia, Rajesh Mahadeva, Vaibhav Gupta, Ebtsam Ahmad Siddiqui, Mamta Patankar, and Vinay Gupta

Abstract Breast cancer is a very commonly diagnosed disease and is a leading cause of cancer deaths in women. Many of the early diagnosis systems have been developed, but the performance of these systems is quite low. Deep learning is an emerging and promising area for medical imaging and other applications, in which classification of breast cancer is a very challenging application. In this work, the cross-level attention (CLA) module is designed to improve the feature gradient in the feature maps to assure the performance improvement of the suggested system. The fusion of all CLA features is concatenated and finally classified by the fully connected neural network. The VGG19 network is considered the base network because of its less complex structure, easy implementation, and faster training. The

V. Patel (✉) · V. Chaurasia · E. A. Siddiqui · M. Patankar
Maulana Azad National Institute of Technology, Bhopal, Madhya Pradesh 462003, India
e-mail: vivekpatel.iet46@gmail.com

V. Chaurasia
e-mail: vijayshree21@gmail.com

E. A. Siddiqui
e-mail: ebtasam.bh27@gmail.com

M. Patankar
e-mail: patankarmamta@gmail.com

R. Mahadeva · V. Gupta
Khalifa University of Science and Technology, Abu Dhabi 127788, United Arab Emirates
e-mail: rajeshmahadeva15@gmail.com

V. Gupta
e-mail: vinay.gupta@ku.ac.ae

R. Mahadeva
Division of Research and Innovation, Uttaranchal University, Dehradun, India

V. Gupta
College of Information Science and Engineering, Ritsumeikan University, Kita-Ku, Kyoto 603-8577, Japan
e-mail: is0591ie@ed.ritsumei.ac.jp

© The Author(s), under exclusive license to Springer Nature Switzerland AG 2025
S. Sambhav et al. (eds.), *Empowering Solutions for Sustainable Future in Science and Technology*, SpringerBriefs in Applied Sciences and Technology,
https://doi.org/10.1007/978-3-031-77837-7_7

proposed model gives 98.04% accuracy, which is 1.07% better in comparison with the other existing methods for breast cancer diagnosis form histopathology images.

Keywords Deep learning · Transfer learning · Breast cancer diagnosis · Cross-level attention · Convolutional neural network

1 Introduction

World Health Organization (WHO) reported that cancer cases have grown within the time span of 2000 to 2020 from 10 million to 19.3 million within 20 years only [1, 2]. Benign and malignant class of labeled histopathology image samples is shown in Fig. 1. To improve the observation of tissues in histopathology image, hematoxylin and eosin (H&E) staining is followed mostly and this is a "gold standard".

Today, computer-aided diagnosis (CAD) is a very reliable and fast process for cancer detection. Many state-of-the-art (SOTA) deep networks have been developed in the last decade, like VGGNets, ResNets, DenseNets, etc. These networks are widely used in medical imaging [3, 4]. ResNets and DenseNets [5, 6]are the deeper and more complex networks in architecture and have the ability to overcome the vanishing gradient problem [7, 8].

Deep transfer learning [9, 10] is a good choice to overcome the training time because a pretrained network considered the base network is used [11, 12]. Regularization technics have the popularity and are used because of robust learning of the deep neural networks, and graph-based method was used in [9]. Thousands of lives can be saved with every percent improvement in the performance of CAD systems [13, 14]. In this paper, a more efficient and compact system is proposed for breast cancer diagnosis. Major contributions to the proposed work are to achieve an edge over the limitations of the existing methods, and a CLA-based deep learning networks is proposed.

Fig. 1 a Benign sample;
b malignant sample

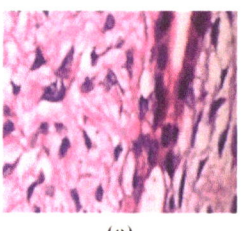 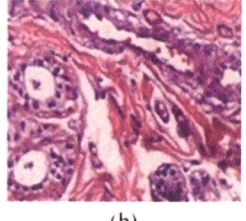

(a)   (b)

2 Methodology

In this work, a CLA-based deep network is proposed for the diagnosis of breast cancer from the histology images. The description of the experimental setup, dataset used, and proposed framework are discussed below.

Experimental Setup and Datasets

Experimental model is implemented with Python 3.9.7, TensorFlow 2.5.0, and Keras 2.4.0 on the dedicated Intel core i7 processor with 2.7 GHz operating speed, 16 GB NVIDIA GeForce GTX 1080 GPU-based 64-bit system. The BreaKHis dataset has 2480 benign and 5429 malignant image samples; 7909 samples overall and are split into a ratio of 7:2:1 for training, testing, and validation sets respectively. Details of both datasets are shown in Table 1, available at https://web.inf.ufpr.br/vri/breast-cancer-database/ and https://iciar2018-challenge.grandchallenge.org/Dataset/. All the images are primarily flipped, resized, and rotated with 30°, 90°, 120°, and 180° to have lager training data.

Proposed Model

Systematic illustration of the proposed network is shown in Fig. 2, which has different submodules such as (1) feature extraction module, (2) cross-level attention module, and (3) classification module. All the submodules are discussed below.

Table 1 BreaKHis and BACH-2018 datasets

Dataset	Class label	Train	Test	Validation	Total
BreaKHis	Benign	1736	496	248	2480
	Malignant	3800	1086	543	5429
BACH-2018	Benign	70	20	10	100
	Malignant	70	20	10	100

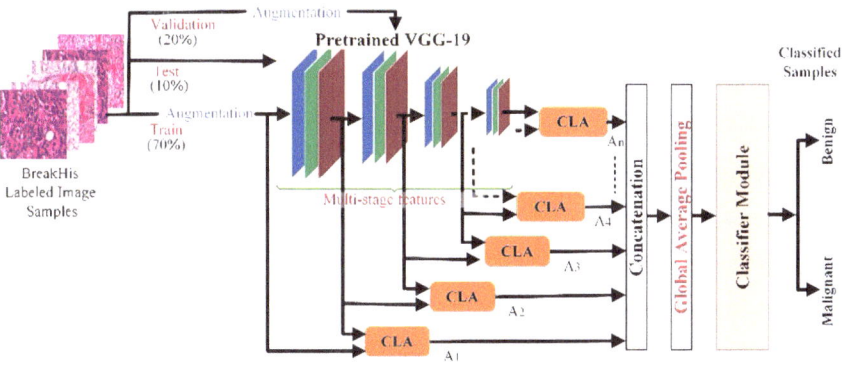

Fig. 2 Proposed cross-level feature attention model for breast cancer histopathology image classification

A. Feature Extraction

Here VGG19 is considered as the base network for feature extraction with TL approach. The input samples are $X^j \in \mathfrak{R}^{M \times M \times C}$, and intermediate stage features from the pretrained backbone network are $F^p \in \mathfrak{R}^{\frac{M}{2k} \times \frac{M}{2k} \times kC'}$. Here M is the size of the input sample ($M = 224$), C is defined as available channels, and k is the number of intermediated stages in the base network.

B. Cross-Level Attention Module

The CLA feature maps are $A_i \in \mathfrak{R}^{\frac{M}{2k} \times \frac{M}{2k} \times kC'}$, where A_i is the ith CLA feature map. The proposed CLA module structure is shown in Fig. 3. The first feature map is the input sample which is resized and reshaped to be compatible with the first interme- diate feature map, and elementwise sum is performed. The bilinear interpolation is used for down sampling, and to equalize the number of channels to the subsequent feature map 1×1 convolution is used. The process of resizing and reshaping is shown in Fig. 4. Mathematically, it can be defined as $Y^p = Con_{1X1}(Biintp(F^p))$ where $p \in (0, 1, \dots n)$ and n is the number of CLA feature maps. Let U^p be the sum of cross-level feature maps as $U^p = F^p \oplus F^{(p+1)}$. Now, the attention feature maps are defined in Eq. (1):

$$A_i = (Sigmoid(PReLU(U^p) \oplus Y^p \tag{1}$$

The VGG19 network is sparse and light weight and has 5 stages within that. Including the input stage 6 stage are considered, i.e., $n=5$. To get gradient rich features, all the CLA features are concatenated: $X_c = concat(A_1, A_2 \dots, A_n)$.

Fig. 3 Cross-level attention (CLA) module; $\oplus$ is depicted for elementwise sum and $\otimes$ is depicted for elementwise multiplication

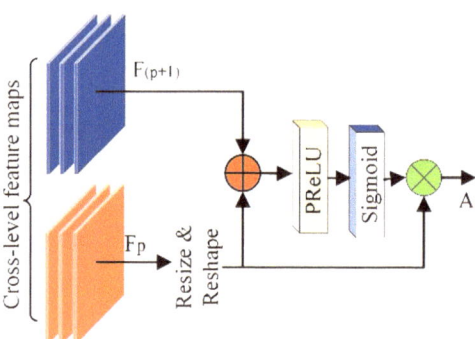

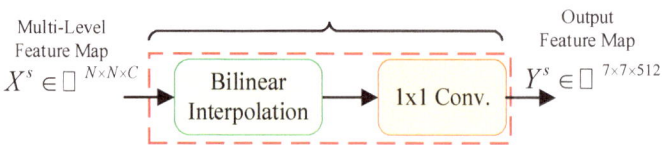

Fig. 4 Resize and reshape process of feature maps

C. Classification Module

Global average pooling (GAP) of all the concatenated CLA feature maps $X_c \in R^{R \times W \times C''}$ is classified by fully connected deep network. Batch Normalization (BN) [9] and dropout layers are included in the classification module to overcome overfitting, while the network is training. Softmax is to predict the sample class, mathematically defined as $(G^i)_{SoftMax} = \hat{y}_j = \frac{\exp(G^i)}{\sum_{J=1}^{s} \exp(G^J)}$. Here G^i is the input feature vector; s is defined as the defined classes; here $s = 2$ for defined two classes, $\hat{y}_j$ is predicted value, and ground truth is defined as y_j. The estimated cross-entropy loss is computed as in Eq. (2) and optimized using the Adam optimizer during training of the network:

$$L = -\frac{1}{2} \sum_{j=1}^{2} y_j \cdot \log(\hat{y}_j) + (1 - y_j) \cdot \log(1 - \hat{y}_j) \tag{2}$$

3 Results and Discussion

The proposed model is trained for 200 epochs. Accuracy and loss plots are shown in Figs. 5 and 6 respectively as the outcome of the proposed network. The confusion matrix after classification of the BreaKHis image samples is shown in Fig. 6. Experimental results are presented in Table 2 with the state-of-the-art methods for performance assessment. Furthermore, experimentation is carried out to diagnose breast cancer from the BACH2018 Kaggle competition dataset. The proposed method is performed over BACH2018 dataset and respective ROC curve is presented in Fig. 7. Figure 8 shows the effectiveness of the technic used. Classification accuracy from the proposed method is 95.00% as presented in Table 3.

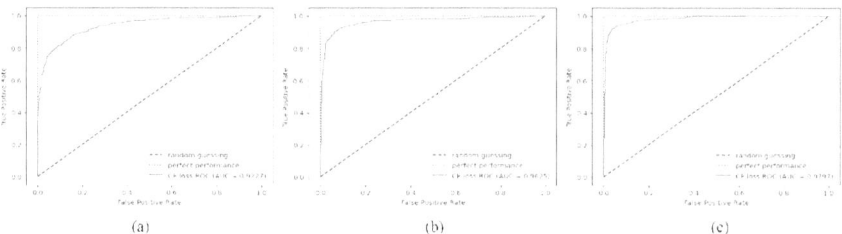

(a)   (b)   (c)

Fig. 5 ROC curve for breast cancer classification from BreaKHis dataset using **a** pretrained VGG19 network and **b** VGG19 with multi-stage feature fusion; **c** proposed (VGG19 with multi-stage feature fusion and cross-level attention)

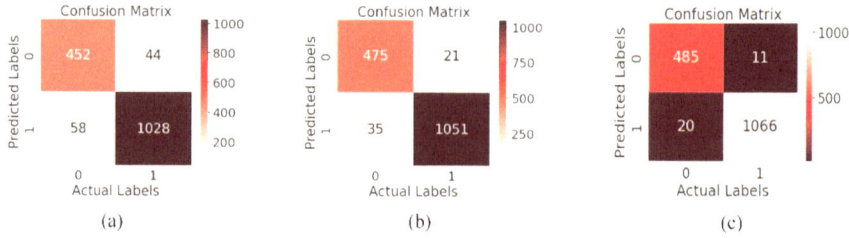

Fig. 6 Confusion matrix for breast cancer classification from BreaKHis dataset using **a** pretrained VGG19 network and **b** VGG19 with multi-stage feature fusion; **c** proposed (VGG19 with multi-stage feature fusion and cross-level attention)

Table 2 Performance comparison of the proposed model with SOTA methods for breast cancer classification from BreaKHis dataset

Dataset	Methods	Acc (%)	Prec (%)	Rec (%)	F1-Score (%)
BreaKHis dataset	[6]	93.06	93.04	94.44	95.28
	[7]	>90.00	97.00	92.00	94.00
	[5]	92.52	93.45	93.23	93.69
	[14]	96.97	–	–	–
	[3]	93.45	–	–	–
	[8]	93.35	–	–	–
	VGG19	93.55	91.13	88.63	89.86
	VGG19 + multi-stage feature	96. 46	95.77	93.14	94.43
	Proposed	98.04	97.78	96.04	96.90

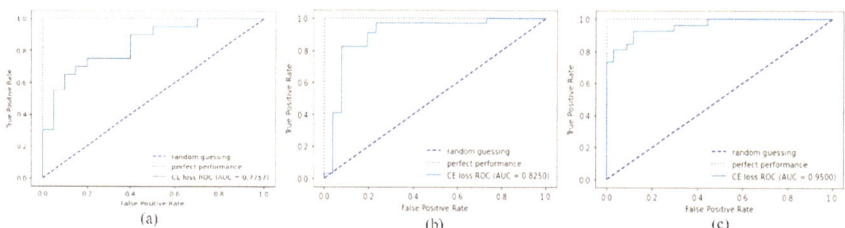

Fig. 7 ROC curve for breast cancer classification from BACH dataset using **a** pretrained VGG19 network and **b** VGG19 with multi-stage feature fusion; **c** proposed (VGG19 with multi-stage feature fusion and cross-level attention)

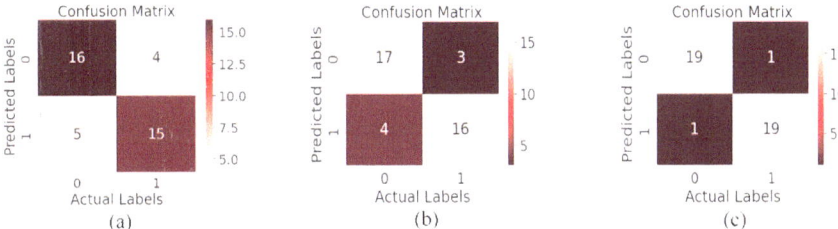

Fig. 8 Confusion matrix for breast cancer classification from BACH dataset using **a** pretrained VGG19 network; **b** VGG19 with multi-stage feature fusion; **c** proposed

Table 3 Performance comparison of the proposed model with SOTA methods for breast cancer classification from BACH2018 dataset

Dataset	Methods	Acc (%)	Prec (%)	Rec (%)	F1-Score (%)
BACH-2018	[15]	85.00	–	–	–
	[10]	76.00	–	–	–
	[11]	87.00	–	–	–
	[4]	92.50	–	–	–
	VGG19	77.50	80.00	76.19	78.05
	VGG19 + multi-stage feature	82.50	85.00	80.95	82.93
	Proposed	95.00	95.00	95.00	95.00

4 Conclusion

The proposed work gives better efficiency over the other works suggested before. The classification accuracy is 98.04% in the classification of breast cancer from BreaKHis images, and 95.00% accuracy for the BACH2018 images. Here a binary classification model is designed in which a sample is classified as benign and malignant class. This work can be further enhanced with the residual features fusion approach to get more efficient and robust diagnosis of breast cancer.

References

1. R.L. Siegel, K.D. Miller, H.E. Fuchs, A. Jemal, Cancer statistics, 2021. CA Cancer J. Clin. **71**(1), 7–33 (2021). https://doi.org/10.3322/caac.21654
2. WHO, Breast cancer key facts, WHO (2023). https://www.who.int/news-room/fact-sheets/det ail/breast-cancer
3. S. Dabeer, M.M. Khan, S. Islam, Cancer diagnosis in histopathological image: CNN based approach. Inform. Med. Unlocked **16**(August), 100231 (2019). https://doi.org/10.1016/j.imu. 2019.100231.

4. Y. Wang, L. Sun, K. Ma, J. Fang, *Breast cancer microscope image classification based on CNN with image deformation*, vol. 10882 (LNCS, Springer International Publishing, 2018). https://doi.org/10.1007/978-3-319-93000-8_96

5. M. Gour, S. Jain, T.S. Kumar, Residual learning-based CNN for breast cancer histopathological image classification. Int. J. Imaging Syst. Technol. **30**(3), 621–635 (2020). https://doi.org/10.1002/ima.22403

6. K. Das, S. Conjeti, J. Chatterjee, D. Sheet, Detection of breast cancer from whole slide histopathological images using deep multiple instance CNN. IEEE Access **8**, 213502–213511 (2020). https://doi.org/10.1109/ACCESS.2020.3040106

7. M.D.P. Yamlome, A.D. Akwaboah, A. Marz, Convolutional neural network based breast cancer histopathology image classification. IEEE Conf. 1144–1147 (2020)

8. R. Kashyap, Dilated residual grooming kernel model for breast cancer detection. Pattern Recognit. Lett. **159**, 157–164 (2022). https://doi.org/10.1016/j.patrec.2022.04.037

9. V. Patel, V. Chaurasia, R. Mahadeva, S.P. Patole, GARL-Net: graph based adaptive regularized learning deep network for breast cancer classification. IEEE Access **11**(January), 9095–9112 (2023). https://doi.org/10.1109/ACCESS.2023.3239671

10. C.A. Ferreira et al., Classification of Breast Cancer Histology Images Through Transfer Learning Using a Pre-trained Inception Resnet V2. Lecture Notes in Computer Science (including subseries Lecture Notes in Artificial Intelligence and Lecture Notes in Bioinformatics) LNCS, 10882, 763–770 (2018). https://doi.org/10.1007/978-3-319-93000-8_86

11. T. Iesmantas, R. Alzbutas, Convolutional capsule network for classification of breast cancer histology images 10882 (LNCS. Springer International Publishing, 2018). https://doi.org/10.1007/978-3-319-93000-8_97

12. R. Mahadeva, S.P. Patole, V. Patel, V. Chaurasia, A. Sharma, R. Sharma, Deep transfer learning with multi-level features extraction approach for breast cancer classification. in 1st IEEE *International Conference Innovations High Speed Communication and Signal Processing IHCSP 2023*, pp. 471–474 (2023). https://doi.org/10.1109/IHCSP56702.2023.10127180

13. V. Patel et al., Breast cancer diagnosis from histopathology images using deep learning methods: a survey. E3S Web Conf. **430**, 1–12 (2023). https://doi.org/10.1051/e3sconf/202343001195

14. Y. Yari, T.V. Nguyen, H.T. Nguyen, Deep learning applied for histological diagnosis of breast cancer. IEEE Access **8**, 162432–162448 (2020). https://doi.org/10.1109/ACCESS.2020.3021557

15. Y. Zou, J. Zhang, S. Huang, B. Liu, Breast cancer histopathological image classification using attention high-order deep network. Int. J. Imaging Syst. Technol. **32**(1), 266–279 (2022). https://doi.org/10.1002/ima.22628

Chaos Analysis in Lowest Dimensional Fractional Order Satellite Systems and Its Control Techniques

Sanjay Kumar, Ram Pravesh Prasad, Mahendra Pratap Pal, Krishan Pal, Ajit Singh, and Vishal Srivastava

Abstract In this chapter, we delve into the exploration of fractional derivatives and chaos analysis within the realm of lowest order chaotic satellite systems. Our investigation extends to the application of fractional calculus in computer simulation, wherein we scrutinize the characteristics of a fractional derivative satellite system through phase portrait analysis across different parameter values. The evolving parameter values give rise to diverse phase portrait analyses of fractional derivatives within the satellite systems, uncovering chaos within systems with fewer than three dimensions. The assessment encompasses equilibrium points, Lyapunov exponents, and bifurcation diagrams as key techniques employed to validate our findings. We have proposed a feedback control strategy for a novel fractional-order satellite system.

Keywords Fractional derivative calculus · Chaos · Satellite systems · Bifurcation theory

S. Kumar (✉) · V. Srivastava
Amity School of Engineering and Technology, Amity University Patna, Patna, Bihar 801503, India
e-mail: sanjay.jmi14@gmail.com

V. Srivastava
e-mail: vsrivastava@ptn.amity.edu

R. P. Prasad · A. Singh
Department of Mathematics, Hansraj College, University of Delhi, Delhi 110007, India
e-mail: ram.mbhudu@gmail.com

A. Singh
e-mail: ajeetsingh@hrc.du.ac.in

M. P. Pal
Department of Mathematics, Sri Venkateswara College, University of Delhi, New Delhi 110021, India
e-mail: mpal@svc.ac.in

K. Pal
Department of Mathematics, Maharaja Agrasen College, University of Delhi, New Delhi 110096, India
e-mail: kpal1987@gmail.com

1   Introduction

Fractional calculus stands out as a premier method for modelling dynamic systems, permeating various scientific, engineering, and interdisciplinary domains. Diverse physical systems, such as the fractional-order Chen system, fractional-order hyper-chaotic Chen system, fractional-order Lorenz system, fractional-order Rossler system, fractional derivative Duffing system, and fractional derivative financial system, have been meticulously characterized through chaotic or hyper-chaotic fractional-order differential equations [1–3]. The application of fractional deriva-tives extends across multiple fields, including medicine, biological tissues, bioengi-neering, ECG testing, cardiac tissue, photoelasticity, fluid mechanics, and materials science [4–6]. Notably, fractional-order and equivalent ordinary differential equation systems diverge in several aspects [7]. In their studies, Kumar et al. employed sliding mode techniques to delve into fractional derivative Rabinovich-Fabrikant systems and fractional derivative hyper-chaotic financial systems [8, 9]. The exploration of chaos control and synchronization in fractional-order systems has garnered signif-icant attention in research. Kumar and Khan, in multiple research works [10–12], have assessed chaos in satellite systems using various tools and established synchro-nization methodologies. However, further study is warranted to comprehensively investigate these systems. In this chapter, we examine the fractional-order chaotic behaviour of satellite systems with various orders, because of the previous debate. Equilibrium points, dissipativity, bifurcation diagrams, and Lyapunov exponents are used to investigate the nature of fractional-order satellite systems. The occurrence of chaos is thereby justified in the fractional-order satellite system's lowest dimen-sion, which is less than three. We also use feedback control approaches to produce chaos control of fractional-order satellite systems. This measurement can help in telecommunications, weather forecasting, GPS systems and earth observations.

2   Basic Concepts of Fractional Derivatives

The non-integer-order integro-differential operator $aD_t{}^{\alpha 1}$ in fractional calculus serves as a generalization of integration and differentiation concepts. It is defined as follows:

$$
_aD_t^{\alpha_1} = \begin{cases} \frac{d^{\alpha_1}}{dt}, & \text{if } \mathcal{R}(\alpha_1) > 0 \\ 1, & \text{if } \mathcal{R}(\alpha_1) = 0 \\ \int\limits_a^t (d\tau)^{-\alpha_1}, & \text{otherwise, } i.e. \ \mathcal{R}(\alpha_1) < 0. \end{cases} \tag{1}
$$

The definition of the generalized Riemann–Liouville definition [5] is given by

$$
D^{\alpha_1}f(t) = \frac{d^{\alpha_1}}{dt^{\alpha_1}}J^{n-\alpha_1}f(t), \ \alpha_1 > 0, \tag{2}
$$

where $n = [\alpha_1]$ and n is the first integer greater than or equal to α_1, and J^{β_1} is the Riemann–Liouville integral operator of order β_1, defined as follows:

$$J^{\beta_1} f(t) = \frac{1}{\Gamma(\beta_1)} \int_0^t \frac{f(\tau)}{(t-\tau)^{1-\beta_1}} d\tau \tag{3}$$

for $0 < \beta_1 \leq 1$, where $\Gamma(.)$ is the gamma function.

The definitions below are utilized as follows:

$$D^{\alpha_1} f(t) = J^{n-\alpha_1} f^n(t), \ \alpha_1 > 0 \tag{4}$$

where $n = [\alpha_1]$. The operator D^{α_1} is commonly referred to as the Caputo differential operator of order α_1 since it was initially employed for the solution of practical problems by Caputo [5].

3　Satellite System Description and Assumptions

The three-dimensional chaotic satellite system is formulated as

$$\dot{x} = \sigma xyz - \frac{1.2}{Ix}x + \frac{\sqrt{6}}{2Ix}z,$$

$$\dot{y} = \sigma yxz + \frac{0.35}{Iy}y, \tag{5}$$

$$\dot{z} = \sigma xyz - \frac{1.2}{Ix}x + \frac{\sqrt{6}}{2Ix}z,$$

with parameters $\sigma x = \frac{Iy-Iz}{Ix}, \sigma y = \frac{Iz-Ix}{Iy}, \sigma z = \frac{Ix-Iy}{Iz}$, and specific values $\sigma x = \frac{1}{3}, \sigma y = -1$, and $\sigma z = 1$. I is the inertia matrix.

The 3D chaotic satellite system becomes

$$\dot{x} = \frac{1}{3}yz - ax + \frac{1}{\sqrt{6}}z,$$

$$\dot{y} = -xz + by, \tag{6}$$

$$\dot{z} = xy - \sqrt{6}x - cz,$$

with known parameters $a = 0.4, b = 0.175$, and $c = 0.4$.

The order $\alpha_1 = 0.95$ of the 3D fractional derivative satellite system is represented as

$$\frac{d^{\alpha_1} x(t)}{dt^\alpha} = \frac{1}{3} yz - ax + \frac{1}{\sqrt{6}} z,$$

$$\frac{d^{\alpha_1} y(t)}{dt^\alpha} = -xz + by, \tag{7}$$

$$\frac{d^{\alpha_1} z(t)}{dt^\alpha} = xy - \sqrt{6}x - cz.$$

Equilibrium Points

The equilibrium points of the satellite system are determined by solving the system

of equations $\dot{X}(t) = 0$. $F(x) = \begin{bmatrix} \frac{1}{3} yz - ax + \frac{1}{\sqrt{6}} z = 0 \\ -xz + by = 0 \\ xy - \sqrt{6}x - cz = 0 \end{bmatrix}$.

Equilibrium Points

$$\aleph_0 = \begin{bmatrix} 0 \\ 0 \\ 0 \end{bmatrix}, \aleph_1 = \begin{bmatrix} 1.1910 \\ 2.5766 \\ 0.3785 \end{bmatrix}, \aleph_2 = \begin{bmatrix} 0.1582 \\ -1.3641 \\ -1.5086 \end{bmatrix},$$

$$\aleph_3 = \begin{bmatrix} -0.1582 \\ -1.3641 \\ 1.5086 \end{bmatrix}, \aleph_4 = \begin{bmatrix} 1.1910 \\ 2.5766 \\ 0.3785 \end{bmatrix}. \tag{8}$$

The satellite system's Jacobian matrix is calculated as

$$J(X) = \begin{bmatrix} -a & -0.33z & \left(0.33y + \frac{1}{\sqrt{6}}\right) \\ -z & b & -x \\ \left(y - \sqrt{6}\right) & x & -c \end{bmatrix} \tag{9}$$

At each equilibrium point, one of the eigenvalues of Jacobian matrix (9) has a positive real portion, indicating that the equilibrium points $\aleph_0$, $\aleph_1$, $\aleph_2$, $\aleph_3$, and $\aleph_4$ are saddle-focus, inherently unstable.

Invariant: y-axis

If $x(0) = 0$ and $z(0) = 0$, then for all t, x and z remain zero according to Eq. (7). As a result, the y-axis is an orbit:

$$\frac{d^{\alpha_1} y(t)}{dt^{\alpha_1}} = by(t), \text{ hence } y(t) = y(0)e^{bt}; \text{ for } x, z = 0. \tag{10}$$

Thus, for the equilibrium, the y-axis illustrates the section of the unstable manifold at the origin.

4 Chaos Control of the Fractional Derivative of Unpredicted Satellite System

The fractional derivative of highly nonlinear system is written as

$$\frac{d^{\alpha_1} x}{dt^{\alpha_1}} = f(x) + \mathbf{B}u \tag{11}$$

where $x \in \mathbf{R}^n$ is a system's state vector, $u \in \mathbf{R}^m$ is a system's input vector, and $f: \mathbf{R}^n \to \mathbf{R}^n$ is a nonlinear map from $\mathbf{R}^n$ to $\mathbf{R}^n$. $\in \mathbf{R}^{nm}$ is a constant matrix. Assume $x^* \in \mathbf{R}^n$ is the controlled chaotic system's equilibrium point. That means, $f(x^*) = 0$.

Remark 1 The goal of this study is to determine and analyse the stability of $\mathbf{x}*$. For the sake of simplicity, we state all formulations and conclusions for the scenario where the equilibrium point of the controlled dynamic system (11) is at the origin of R^n. That is, $\mathbf{x}* = \mathbf{0}$. By adjusting variables, it is possible to move every equilibrium point to the origin; hence there is no loss of generality.

Remark 2 Consider the change of variables $y = x - x^*$ and assume $x^* = 0$ is an equilibrium point of the controlled chaotic system (11). Then, y-derivative is equal to

$$\frac{d^{\alpha_1} y}{dt^{\alpha_1}} = \frac{d^{\alpha_1} x^*}{dt^{\alpha_1}} = f(x) + \mathbf{B}u = f(y + x^*) = s(y), \tag{12}$$

where $s(0) = 0$. The system is in equilibrium at the origin with the new variable y. As a result, we will always assume that $f(x)$ fulfils $f(0) = 0$ and analyse the stability of the origin $x = 0$ without losing generality. Using Taylor series for the nonlinear function $f(x)$ of system (11)

$$f(x) = f(0) + \frac{\partial f(x)}{\partial x} + h(x), \tag{13}$$

where $h(x)$ fulfils $\lim_{\|x\| \to 0} \frac{\|h(x)\|}{\|x\|} = 0$.

We notice that when $f(0) = 0$ is substituted into the controlled chaotic system (11) and combined with (11), we get

$$\frac{d^{\alpha_1} x}{dt^{\alpha_1}} = Ax + h(x) + Bu, \tag{14}$$

where $A = \frac{\partial f(x)}{\partial x}$ at $x = 0$.

4.1 Controllers with Linear State Feedback

Theorem *(A; B) is entirely state controllable for the linearization of dynamical system (14) of controlled chaotic system (11). Control law of the linear state feedback is therefore specified as $u = B^T PX$, where P is a unique positive definite symmetry matrix, and the controlled chaotic system (11) is asymptotically stable in the origin. The equation for Riccati algebra matrices is $PA + A^T P - PBB^T P + Q = 0$, where Q is an arbitrary positive definite matrix.*

4.2 Control of a Fractional Derivative Chaotic Satellite System

The eigenvalues of system (9) at equilibrium point $\aleph_0 = (0,0,0)$ are $\lambda_{01} = 0.175$, $\lambda_{02} = -0.4 + 0.99\imath$, and $\lambda_{03} = -0.4 + 0.99l$ are obtained, where l as $\sqrt{-1}$. Because one of the eigenvalues is positive, both the linearized and original forms of the fractional-order satellite system are unstable. The linear state feedback law of the unpredictable system is provided by after controllers arithmetic,

$$u = -\mathbf{B}^T P\big[(x, y, z)^T\big].$$

5 Numerical Simulation

We initialize the fractional derivative master system with $x(0) = (3.5, 0.5, 0.8)^T$ and the fractional derivative slave satellite system with $y(0) = (2.5, 1.5, 0.3)^T$. Figure 1 presents the phase portraits and time series graphs of the fractional-order satellite system. The characteristics of this system analyses 3D phase portraits and time series diagrams. In Fig. 2, the phase portraits of the three- and two-dimensional fractional-order satellite system are depicted for the fixed order $\alpha_1 = 0.95$. Figure 3a–c shows the bifurcation plots concerning the parameters *a, b,* and *c,* respectively. The Lyapunov exponents of the satellite system at $t = 200$ are calculated and presented in Fig. 4. Feedback control of the unpredictable fractional derivative satellite is illustrated in Fig. 5, initiated at $t = 20$.

6 Conclusions

In this study, we have laid the groundwork for understanding fractional dynamics and delved into the chaotic behaviour exhibited by specific fractional-order satellite systems through the examination of equilibrium points. Various analytical

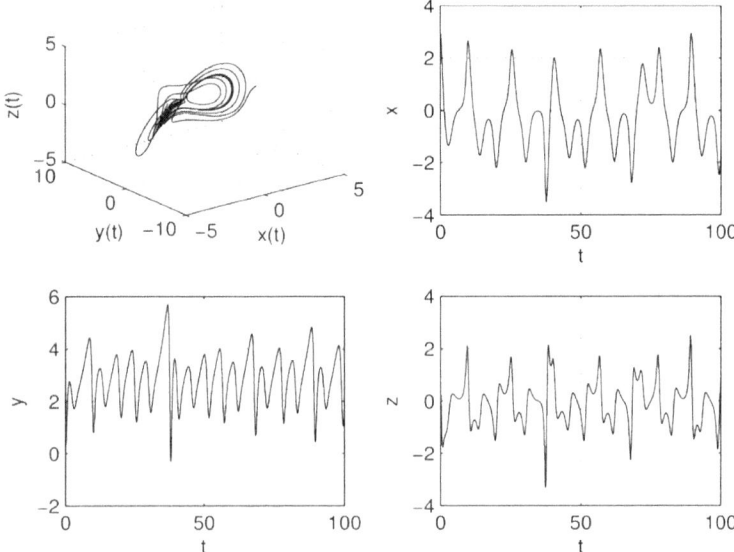

Fig. 1 3D phase portrait and time series graphs of chaotic fractional-order satellite system

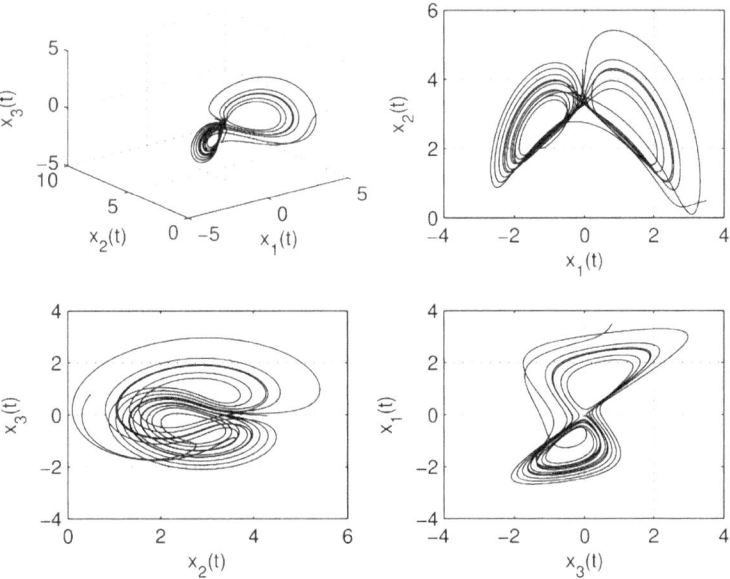

Fig. 2 3D and 2D (dimensional) phase portraits of fractional-order satellite systems with $\alpha_1 = 0.95$

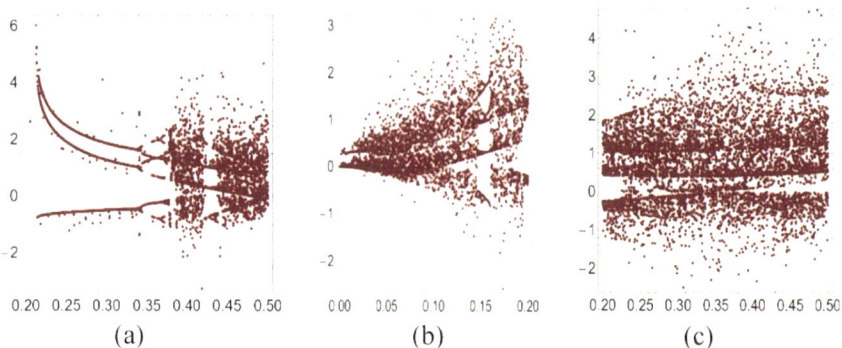

(a)   (b)   (c)

Fig. 3 Bifurcation plotting with varying parameters a, b, and c respectively

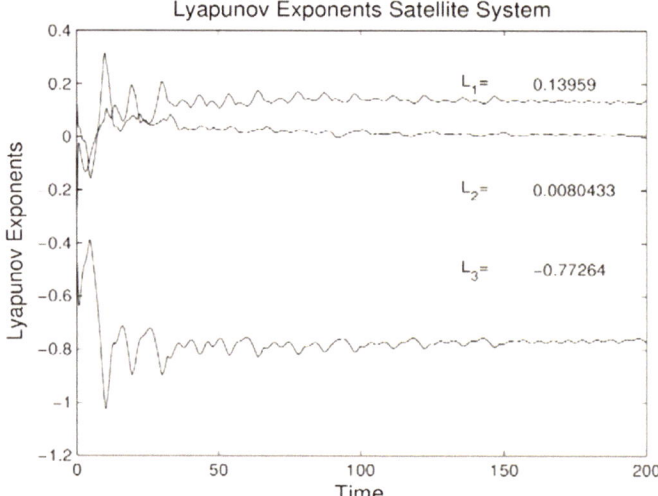

Fig. 4 Lyapunov exponent value of fractional-order chaotic satellite system

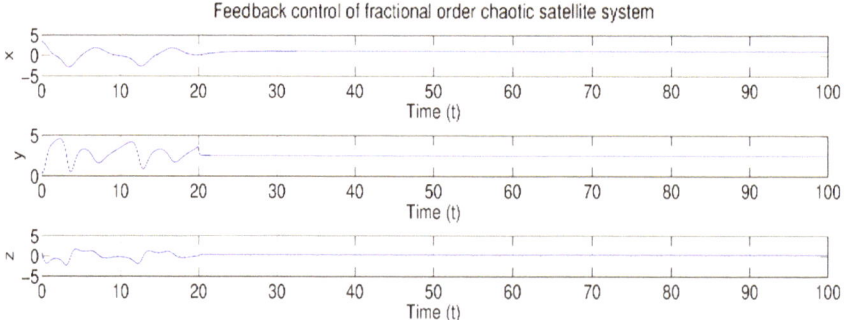

Fig. 5 Feedback controllers of fractional-order chaotic satellite system started at $t = 20$ s

methods, including equilibrium points, bifurcation diagrams, and Lyapunov exponents, have been employed to scrutinize the chaotic dynamics of fractional-order satellite systems, offering insights through phase portrait analysis across different parameter values. The phase portrait analysis of fractional derivatives for diverse satellite systems has been systematically presented. Additionally, we proposed a feedback control strategy for a novel fractional-order satellite system. These investigations hold significant implications for diverse applications, including secure telecommunications, data security, radar detection, weather forecasting, GPS systems, and earth observation, providing a valuable contribution to the fields of technologies and sciences.

References

1. C. Li, G. Chen, Chaos and hyper chaos in the fractional order rossler equations. Physica A **341**, 55–61 (2004)
2. Z. Wang, X. Huang, G.D. Shi, Analysis of nonlinear dynamics and chaos in a fractional order financial system with time delay. Comput. Math. Appl. **62**, 1531–1539 (2011)
3. A.E. Matouk, Stability conditions, hyper chaos and control in a novel fractional order hyperchaotic system. Phys. Lett. A **373**, 2166–2173 (2009)
4. K.P. Wilkie, C.S. Drapaca, S. Sivaloganathan, A nonlinear viscoelastic fractional derivative model of infant hydro-cephalus. Appl. Math. Comput. **217**, 8693–8704 (2011)
5. I. Podlubny, *Fractional Differential Equations* (Academic Press, San Diego (CA), 1999)
6. S. Das, P.A. Gupta, Mathematical model on fractional lotka-volterra equations. J. Theor. Biol. **277**(1), 1–6 (2011)
7. L.M. Pecora, T.L. Carroll, Synchronization in chaotic systems. Phys. Rev. Lett. **64**, 821–824 (1990). E. Ott, C. Grebogi, J.A. Yorke, Controlling Chaos. Phys. Rev. Lett. **64**, 1196–1199 (1990)
8. A. Khan, S. Kumar, Synchronization of fractional-order hyperchaotic finance systems using sliding mode control techniques. IGI Glob. 133–151 (2020)
9. A. Khan, S. Kumar, Synchronization of fractional order Rabinovich-Fabrikant systems using sliding mode control techniques. Arch. Control. Sci. **29**(2) (2019)
10. A. Khan, S. Kumar, Study of chaos in satellite system. Pramana J. Phys. **90**, 13 (2018)
11. A. Khan, S. Kumar, Measuring chaos and synchronization of chaotic satellite systems using sliding mode control. Optim. Control. Appl. Methods (2018). https://doi.org/10.1002/oca.2428
12. A. Khan, S. Kumar, Analysis and time-delay synchronisation of chaotic satellite systems. Pramana J. Phys (2018). https://doi.org/10.1007/s12043-018-1610-5

Harnessing Science for Power: An Analysis of MCU Characters and the Role of Technology in Shaping Power Distribution

Roli Mishra, Arundhati Sharma, Rahul Sharma, Upanshu Mishra, Shatakshi Srivastava, and Shwet Nisha

Abstract The Marvel Cinematic Universe (MCU) stands as a sprawling canvas where science and technology intertwine with the realms of digital humanism and cultural theory. This abstract delves into the intricate web of digital advancements and cultural nuances in the MCU, exploring how science and technology are portrayed and experienced within this cinematic universe. Through the lens of digital humanism, the MCU's portrayal of artificial intelligence, cybernetics, and virtual realities is examined, unraveling ethical questions about human–machine interactions and individual agency. Cultural theory comes into play as the MCU seamlessly weaves diverse cultural elements into its technological landscape, showcasing the complexities of cultural identities and power dynamics. The abstract analyzes iconic instances where technology impacts social structures, identity formation, and cultural representation, shedding light on the MCU's role in shaping contemporary narratives about human-technology relationships. By exploring these themes, this study illuminates the profound impact of science and technology on the human experience, providing insights into the intersection of digital humanism, cultural theory, and the imaginative realms of the MCU.

R. Mishra (✉) · A. Sharma · R. Sharma · U. Mishra · S. Srivastava · S. Nisha
Amity University Patna, Patna, Bihar 801503, India
e-mail: 1323rolimishra@gmail.com

A. Sharma
e-mail: arundhatisharma2092@gmail.com

R. Sharma
e-mail: rahulsharma.kv1987@gmail.com

U. Mishra
e-mail: umishra@ptn.amity.edu

S. Srivastava
e-mail: shatakshi2012@gmail.com

S. Nisha
e-mail: shwetnisha@gmail.com

© The Author(s), under exclusive license to Springer Nature Switzerland AG 2025   73
S. Sambhav et al. (eds.), *Empowering Solutions for Sustainable Future in Science and Technology*, SpringerBriefs in Applied Sciences and Technology,
https://doi.org/10.1007/978-3-031-77837-7_9

Keywords Fiction · Science and technology · Culture · MCU · Marvel

1 Introduction

The Marvel Cinematic Universe (MCU) is a multifaceted and expansive franchise that transcends traditional boundaries within the entertainment industry. Comprising a diverse array of movies, television shows, and comics, the MCU has achieved unprecedented prominence by skillfully integrating elements of science and technology into its overarching narrative [1]. This integration extends far beyond mere storytelling, as it permeates the very essence of the MCU's character development, plot progression, and world-building. The MCU's unique appeal lies in its ability to engage viewers on multiple levels, not only by showcasing dazzling visual effects and thrilling action sequences but also by delving into the intricate relationship between science, technology, and human potential. From its inception, the MCU has adopted a captivating blend of real-world science, speculative fictional technology, and superhuman abilities, all of which coalesce to create a narrative universe that is both thought-provoking and awe-inspiring [2]. The infusion of real-world scientific principles and speculative technology resonates with audiences, who often find themselves pondering the boundaries between what is scientifically plausible and the bounds of pure imagination. This convergence of possibilities and creativity is where cultural theory takes center stage, as it allows us to dissect how the MCU functions as a cultural artifact within the context of the twenty-first century.

Through the lens of cultural theory, the analysis dissects the meticulously crafted portrayals of cultural identities, power structures, and sociopolitical dynamics within the MCU. Characters like T'Challa (Black Panther), embodying Afrofuturism and postcolonial identity, and the inhabitants of Asgard, representing a mythologically rich heritage, emerge as pivotal case studies. By drawing on Hall's concept of cultural representation, the paper critically examines how the MCU navigates themes of heritage, colonialism, and cultural diversity [3]. Hall's ideas on representation as a process of constructing meaning provide a theoretical scaffold to understand how these characters and their narratives challenge prevailing cultural norms while negotiating complex sociopolitical contexts. By scrutinizing these narratives, the paper uncovers the intricate ways in which the MCU both challenges and reinforces cultural stereotypes, fostering a profound dialogue on the nuances of identity within the realm of science fiction. This analysis not only enriches our comprehension of cultural representation in popular media but also showcases the MCU's role as a transformative cultural space, shaping and reshaping narratives about diversity, heritage, and power in contemporary society.

The MCU serves as a cultural touchstone, reflecting the evolving ethos of our time, as it navigates issues such as personal responsibility, the consequences of unchecked power, and the need for collective action. The Marvel Cinematic Universe stands as a prime example of how cultural theory can illuminate the intricate dance between science, technology, and societal values within the realm of popular culture. It offers a platform where audiences are not only entertained but are also encouraged to grapple with profound questions about the past, present, and future of our ever-changing world, creating a rich tapestry that weaves the realms of imagination and reality into a compelling narrative that resonates across the globe. This study offers a thorough exploration of the MCU's technological landscape, highlighting its transformative impact on cultural narratives and fostering discussions on technology, identity, and ethics in our global society [4].

2 Methodology

2.1 Digital Humanism in the MCU: Navigating Ethical Frontiers

In the sprawling saga of the Marvel Cinematic Universe, a subtle yet powerful current of digital humanism runs deep, threading through the fabric of its rich storylines and character arcs [5]. This concept, championed by thinkers like Luciano Floridi and Donna Haraway, is not just a backdrop but a pivotal axis around which the tales of characters such as Tony Stark and Vision revolve, highlighting a future where technology progresses in harmony with human values, dignity, and freedom.

Tony Stark, the man behind Iron Man, is a vibrant illustration of this harmony. His transformation from a merchant of death to a champion for peace reflects the ethical quandaries that accompany the march of technological innovation. As he forges sentient AI and wields digital tools of immense power, Stark embodies the very essence of digital humanism, standing at the juncture where technological advancement must be weighed against its ramifications for society.

Vision, a being of silicon and synthetics, yet suffused with consciousness, brings the philosophy of digital humanism into sharper focus. His synthetic life raises probing questions about the ethical treatment of artificial intelligence within our moral compass. The quandaries of his existence compel us to ponder the very fabric of humanity and the place of AI within it, challenging us to redefine what it means to be human in an age where the lines between organic and artificial life are increasingly blurred [6].

This theme is further explored through the lens of Haraway's cyborg theory, particularly in characters like Bucky Barnes and Nebula, whose very identities are a tapestry of technology and flesh. Their narratives grapple with a posthuman condition, where the cyborg becomes a metaphor for the evolving human identity, challenging age-old dichotomies and signaling a new paradigm of being [7].

The forays into the realms of virtual and augmented reality, such as those depicted in the "Doctor Strange" films and the enigmatic Quantum Realm, broaden the MCU's ethical discourse. Here, characters navigate a liminal space where Floridi's ethical

concerns come to the fore, questioning the authenticity of experience and the consequences of actions within realms that defy the conventional boundaries of physical
and digital existence.

2.2   Cultural Theory in the MCU: Interrogating Identity and Representation

At the heart of the Marvel Cinematic Universe lies a profound intersection of science,
technology, and culture, reflecting the collective human imagination and the societal
context in which it is embedded. Cultural theorists like Marshall McLuhan and
Raymond Williams have highlighted the critical role of popular culture in shaping
contemporary society, and the MCU exemplifies this phenomenon [8]. The MCU's
success is rooted in its ability to harness and explore the theme of superhuman
abilities within a scientifically and technologically grounded context. This approach
has allowed the franchise to speak to a wide and diverse audience, resonating with
viewers who seek a bridge between the extraordinary and the plausible.

Characters such as Tony Stark (Iron Man) and Steve Rogers (Captain America)
serve as prime examples of how science and technology coalesce to create compelling
narratives. Stark's Iron Man suit is a testament to the potential of engineering, pushing
the boundaries of what is achievable through human ingenuity. It symbolizes the
fusion of cutting-edge technology and individual genius, reflecting a cultural fascination with the idea of the brilliant inventor. Steve Rogers' transformation into Captain
America, on the other hand, underscores the role of scientific experimentation. His
journey mirrors the human desire to overcome limitations and become something
greater, a theme that echoes throughout history and literature.

Additionally, the introduction of futuristic technologies like Wakandan Vibra-
nium and the Asgardian Bifrost transcends known scientific principles, crafting a
fantastical world that captures the collective human imagination. This showcases the
power of popular culture, exemplifying how narratives can inspire discussions on the
potential of science and technology in shaping the future [9]. The MCU serves as
a reflection of our fascination with the limitless possibilities that science and technology can offer, while simultaneously reminding us of the ethical dilemmas and
moral quandaries that come with such advancements.

In essence, the MCU's narrative is not merely a vehicle for action and entertainment; it is a cultural artifact that sparks dialogues and reflections on our ever-evolving
relationship with science, technology, and the boundless horizons of human potential.
McLuhan's concept of the "global village" and Williams' idea of "cultural materialism" are encapsulated within the MCU, underscoring how this franchise mirrors
and influences our contemporary cultural landscape, fusing the realms of science and
technology with the collective human imagination [10].

2.3   Science and Technology Used by Characters in MCU and Power Digression

The Marvel Cinematic Universe showcases characters such as Tony Stark, Spider-Man, Bruce Banner, and Shuri, who exemplify the complex relationship between science, technology, and cultural representation.

On the other hand, Spider-Man epitomizes the archetype of the youthful genius whose technological prowess aligns with the millennial generation's integration of tech into traditional heroic tales. Peter Parker's character redefines the potential of science for societal good through the lens of youth.

Bruce Banner's transformation into the Hulk represents the ethical complexities that can accompany scientific discovery, embodying cultural theory discussions on the moral implications inherent in the progression of science.

Shuri, representing the technologically advanced nation of Wakanda, brings Afro-futurism to the forefront, symbolizing a future where technological innovation is synonymous with cultural diversity and inclusivity. Her character reframes the narrative of science and technology within the context of cultural identity.

The utilization of science and technology by these characters not only drives their narratives but also connects with larger discussions within cultural theory regarding science, ethics, and identity.

2.4   Intersections and Impact: Unraveling Cultural Theory in the MCU's Technological Landscape

Stuart Hall's concept of cultural identity as a site of struggle becomes paramount in understanding how characters like T'Challa and Vision navigate the complex interplay of technology, identity, and power. The paper explores how the MCU's portrayal of digital humanism, rooted in ethical dilemmas and human–machine interactions, echoes Hall's ideas of identity negotiation within cultural contexts [11]. Furthermore, Bell Hooks' feminist perspectives offer critical insights into the representation of gender within the MCU, shedding light on how the intersection of technology and cultural identity extends to gender dynamics [12]. Through this lens, the paper unpacks the portrayal of female characters like Natasha Romanoff (Black Widow), examining how their narratives challenge traditional gender roles and contribute to evolving feminist discourses.

The MCU's commitment to scientific and technological themes has sparked a groundswell of discussions, fan theories, and educational initiatives. The franchise's popularity acts as a catalyst for constructive dialogues within popular culture, fueling debates on subjects ranging from quantum physics to ethical questions surrounding artificial intelligence. This aligns with Henry Jenkins' notion of participatory culture, where fans become active contributors to the development of content, generating fan theories that not only enhance the viewing experience but also encourage a deeper exploration of scientific concepts.

Furthermore, the MCU's influence extends to science, technology, engineering, and mathematics (STEM) education initiatives. The franchise serves as an engaging and relatable vehicle to pique the curiosity of young minds about the sciences. As they watch characters like Tony Stark or Shuri (from Black Panther) demonstrate the marvels of technology, they are inspired to delve into real-world STEM subjects, driven by the desire to understand the principles behind the fictional gadgets and innovations portrayed on screen. This connection between the MCU and STEM education aligns with contemporary efforts to make science and technology more accessible and engaging for future generations.

3   Results and Discussion

A survey was conducted by means of a questionnaire circulated among a heterogeneous group.

The survey aimed to find out which among the following characters in the MCU universe is more tech-savvy or technological prodigy, Figs. 1 and 2.

Shuri was declared the one with the maximum tech-savvy with 38%, followed by Iron Man, Bruce Banner, and Spider Man getting 30%, 22%, and 10% respectively.

The following data suggests the answer of the questionnaire as per Fig. 3.

Fig. 1 Shuri dismantling vision [13]

Fig. 2 Iron man in iron man 2, building the ultron [14]

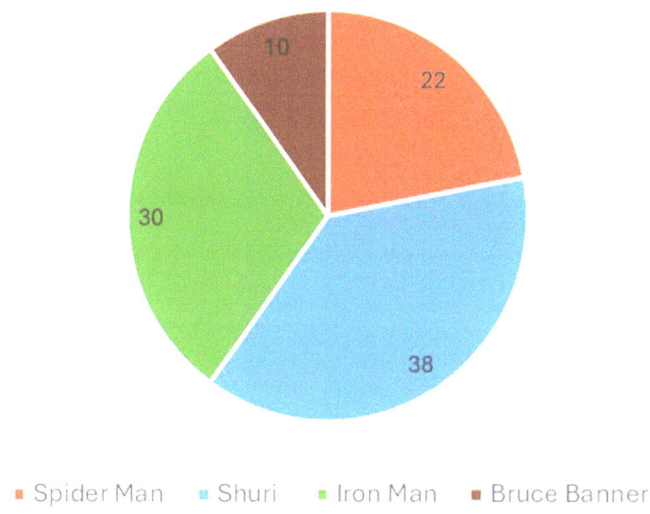

Fig. 3 Proportion of tech-savvy characters in MCU

4 Conclusion

Traversing the intricate realms of digital humanism and cultural theory within the Marvel Cinematic Universe, this research paper unveils a multifaceted tapestry that extends beyond mere entertainment. It serves as a powerful testament to the profound societal impacts of technology, ethics, and cultural identity. The MCU, functioning as both a cultural mirror and a transformative space, masterfully captures the symbiotic relationship between these themes. The application of cultural theory, particularly in the context of characters like T'Challa and the inhabitants of Asgard, provides a captivating lens through which we discern the nuanced portrayal of heritage, colonialism, and diversity. Simultaneously, the exploration of digital humanism, exemplified through the journeys of characters like Tony Stark and Vision, plunges into the ethical dilemmas born from advanced technology.

Through its captivating characters like T'Challa and the denizens of Asgard, the MCU becomes a vibrant reflection of the complexities of heritage, colonialism, and diversity, all the while challenging prevailing cultural stereotypes. It offers a transformative space where identity formation is explored, reshaping societal perceptions, and underscoring the potential of popular media to be both a mirror and a transformative catalyst.

The MCU's utilization of science and technology to maintain power distribution offers a multi-layered exploration of power dynamics within popular culture. Characters like Tony Stark, Shuri, and Bruce Banner serve as vehicles for examining the intersections of science, culture, and power, as theorized by Raymond Williams and Michel Foucault. The MCU's narratives provide an insightful commentary on the complexities of scientific advancement.

References

1. A.J. Friedenthal, *The World of Marvel Comics,* (Routledge, 2021)
2. A. Gibson, Book review: biology in the marvel cinematic universe. J. Hist. Biol. **52**(2), 365–369 (2019). https://doi.org/10.1007/s10739-019-9564-0
3. S. Hall, P. Du Gay, *Questions of Cultural Identity,* (SAGE Publications SAGE, 1996)
4. T. De Luca, *Slow Cinema,* (Edinburgh University Press, 2015)
5. J.C. Taylor, Reading the marvel cinematic universe: the avengers' intertextual aesthetic. JCMS **60**(3), 129–156 (2021). https://doi.org/10.1353/cj.2021.0030
6. B. Davis, *Movie comics: Page to Screen/screen to Page,* (2017)
7. A. Verbeke, R. Van Tulder, E.L. Rose, Y. Wei, *The multiple dimensions of institutional complexity in international business research,* (Emerald Group Publishing, 2021)
8. R. Williams, *Culture and materialism,* (Verso, 2005)
9. A. Heys, *The anatomy of bloom: Harold Bloom and the Study of Influence and Anxiety,* (Bloomsbury Publishing USA, 2014)
10. M. McLuhan, *The Gutenberg Galaxy,* (Signet, 1969)
11. S. Russell, P. Norvig, Artificial intelligence: a modern approach. Choice Rev. Online **33**(03), 33–1577 (1995). https://doi.org/10.5860/choice.33-1577
12. S. Kidwai, The use of language in bell hooks' book 'feminism is for everybody: passionate politics. Int. J. Adv. Res. Manag. Soc. Sci. **4**(11), 204–211 (2015). https://www.garph.co.uk/IJARMSS/Nov2015/21.pdf
13. "Why does the Mind Stone still affect Vision after Shuri spent all that time separating it?," *Movies & TV Stack Exchange.* https://movies.stackexchange.com/questions/93244/why-does-the-mind-stone-still-affect-vision-after-shuri-spent-all-that-time-sepa
14. V. Frei, Iron Man 2: Danny yount—Creative director—Prologue films—The art of VFX. Art VFX (2014). https://www.artofvfx.com/iron-man-2-danny-yount-directeur-artistique-prologue/

Empowering Anaphora Resolution in Magahi Through Decision Trees

Shweta Chandra and Lalita Kumari

Abstract One of the fundamental issues in natural language processing (NLP) is anaphora resolution, which is matching pronouns such as 'he', 'she', 'it', or 'they' to the right antecedents or referents in a text. Finding the words or phrases in the text that these pronouns are referring to is the main objective of anaphora resolution. Pronouns are frequently ambiguous, and several words or phrases in the text may potentially act as their antecedents, making this problem difficult to solve. To properly assign each pronoun to the appropriate referent, anaphora resolution necessitates taking the text's semantics and context into account. Many NLP applications, such as machine translation, text summarization, question answering, and co-reference analysis, depend on successful anaphora resolution. The results demonstrate the effectiveness of our proposed methodology in resolving anaphoric references in Magahi texts.

Keywords Anaphora resolution · Referring expressions · Pronouns and ambiguity

1 Introduction

One of the fundamental issues in natural language processing (NLP) is anaphora resolution, which is matching pronouns such as 'he', 'she', 'it', or 'they' to the right antecedents or referents in a text. Finding the words or phrases in the text that these pronouns are referring to is the main objective of anaphora resolution. Pronouns are frequently ambiguous, and several words or phrases in the text may potentially act as their antecedents, making this problem difficult to solve. To properly assign each pronoun to the appropriate referent, anaphora resolution necessitates taking the text's semantics and context into account. Many NLP applications, such

S. Chandra
Amity School of Civil Services, Amity University Patna, Patna, Bihar 801503, India
e-mail: schandra@ptn.amity.edu

L. Kumari (✉)
Amity School of Engineering and Technology, Amity University Patna, Patna, Bihar 801503, India
e-mail: lkumari@ptn.amity.edu

as machine translation, text summarization, question answering, and co-reference analysis, depend on successful anaphora resolution. It plays a vital role in enabling machines to understand and interpret natural language texts, making it an essential component of NLP research and applications.

To connect pronouns and other referring expressions to their intended referents in text, anaphora resolution solves a difficult problem in natural language processing. Because it allows machines to effectively understand and interpret human language, this task is extremely important.

Data collection: Given the language's low online presence in the context of this study, gathering data for the resolution of anaphora in the Magahi language posed substantial problems. There were several reasons why substantial manual data gathering was necessary, including the rarity of Magahi text passages, the absence of annotated datasets, domain-specific data requirements, and ethical worries. The difficulties highlight the necessity of tailored approaches for collecting data when access to language- or domain-specific information is limited.

Related Work:

Research on Indian languages is scarce, even though anaphora resolution in English has been thoroughly examined and numerous strategies suggested. Languages can be classified into two categories: verbal and nominal devices, or both. Simplex or complex anaphors make up the nominal device in Indo-Aryan languages. Both nominal and verbal methods are used in Manipuri. While verbal reciprocal is required in Munda, Tibeto-Burman, and Dravidian, the verbal reflexive morpheme is optional in Kannada. The Austro-Asiatic, Dravidian, Indo-Aryan, and Tibeto-Burman language families have been the subject of theoretical studies on anaphora [1]. Conversation translation and ambiguity in machine translation situations have been the focus of computational research. The purpose of this study is to identify the features of Magahi's grammatical patterns. From a technical standpoint, co-referencing can be represented as a hyperlink inside natural language that primarily links a stylistic and cohesive aspect to language speakers or writers and co-relates an additional level of obscurity to the mechanical comprehension of language [2, 3]. Comprehensive studies in this and the related field of anaphora resolution have been conducted in the past. Research on anaphora can often be categorized into two main groups when it comes to literature reviews. It is evident that the first section is linguistically motivated, or rule based, mostly based on syntax, morphology, focus, and centering theory, and it largely relies on language and domain knowledge [4, 5]. The application of machine learning techniques—such as conditional random fields, co-training, decision trees, clustering, and others—to co-reference resolution is the focus of the second section. These techniques rely on data-driven methodologies. There has, however, also been some meager research done on the use of hybrid approaches (i.e., rule based and machine learning based) [6].

Thus, pronoun anaphora and other closely similar anaphora are the key topics of discussion for linguistic approaches within the first classification.

i. **Machine Learning Approaches**: This section discusses machine learning approaches, focusing on co-reference resolution since the mid-to-late 90 s.

The machine learning community has developed various techniques, including statistical naïve Bayes models, conditional random fields, and decision trees.

ii. **Naive Bayes' Model**: The naive Bayes' model proposed a statistical technique for anaphora resolution influenced by syntactic and semantic factors [7]. The model considers distance, syntactic structure, gender, number, and animacy as factors. The mention count is motivated by centering theory, which suggests a continuous topic is the highest ranked candidate for a pronoun. A modified version of Hobbs' algorithm is used to compute the distance between antecedents and pronouns.

iii. **Decision Trees**: To resolve co-references, decision trees are a classification technique that concentrates on whether markable types (namely, demonstrative noun phrase, pronoun, reflexive pronoun, definite noun phrase, and proper noun) are related to one other. Using a decision tree technique, this method finds every potential markable type during preprocessing stages and learns rules based on various attributes computed on pairs of markable types. Markable types, agreement, distance, semantic class agreement, and alias features are important aspects. Positive training samples are produced automatically from the corpus using nearby noun phrase pairs, whereas negative training examples are produced using non-referent pairs. The feature is now set with new features and separated pronominal and non-pronominal nouns to modify the framework [8]. Using two sets of positive and negative candidate antecedents and anaphora for training, [9, 10] developed a competition-based learning strategy. The precise setup and decision tree employed in the twin candidate model are not evident, but they aid in distinguishing between a positive and negative candidate for an anaphor.

iv. **Conditional Random Fields (CRFs)**: To overcome co-reference concerns, three types of conditional random fields were suggested [11]. In the first model, conditional on entity mentions, co-reference decisions, and entity attributes are treated as random variables. In the second model, a binary valued random variable takes the place of the co-reference variable, eliminating its reliance. With its exclusion of characteristics as a random variable, the third model outperforms the second by a small margin. CRFs can manage inference issues such as graph partitioning and transitive dependencies. Wellner et al. suggested a framework for named entity extraction and co-resolution of proper names, enhanced approach, and skip-chain CRFs for proper name co-reference are some more approaches [12]. However, it is unclear if an integrated model will be useful for resolution of other types of noun phrases.

2 Methodology

In this research, we propose an algorithm for anaphora resolution using a neural network. The algorithm consists of several key steps—(a) Data preparation: The initial step involves the loading and preprocessing of a dataset containing text passages with anaphoric pronouns and their respective antecedents. Labels are assigned to indicate the locations or labels of the antecedents for each pronoun. In the data preprocessing stage, the following tasks were conducted: (i) Sentence tokenization—the Magahi text was initially tokenized into sentences using the NLTK library. Each sentence was considered as a separate unit for analysis. (ii) Word tokenization—within each sentence, word tokenization was performed to segment the text into individual words or subword units. This step enabled a more granular analysis of the text. (iii) Anaphoric reference extraction—anaphoric references, specifically pronouns and their corresponding antecedents, were identified within each tokenized sentence. Pronouns, including 'je', 'ham', 'ekar', and 'jekara', were singled out for analysis. (iv) Antecedent identification—the antecedents of these pronouns were determined by examining the word immediately preceding the pronoun in each sentence. (v) Dictionary representation—the identified anaphoric references were stored in a dictionary, linking each pronoun to its antecedent. This dictionary format made it easier to access and manage the anaphoric references.

i. **Model Selection (Neural Network Model)**: Neural network architecture is defined catering to the specific requirements of the anaphora resolution task. The choice of the neural network model, such as a feedforward neural network, RNN, LSTM, and transformer-based model, is made based on the problem's nature and complexity.

ii. **Training**: The model is initialized with random weights. A suitable loss function and optimization algorithm are selected. The training process involves iterating over the training dataset. During each iteration, the forward pass is performed, where text passages are fed through the model to make predictions. The loss is calculated, representing the difference between predicted labels and the ground truth. Error is back-propagated, updating the model's weights to improve performance.

In this paper, textual data undergoes preprocessing using Natural Language Toolkit (NLTK) for tasks like tokenization, stemming, and stop-word removal. These processed texts serve as input to the RNN. A dedicated anaphora resolution neural network is designed, specifically employing RNN architecture for the task. Unlike BERT, which is transformer-based, RNNs are recurrent models. The RNN-based neural network is trained on an annotated dataset, which consists of pairs of pronouns and their corresponding antecedents. During training, the primary goal is to teach the model to predict the correct antecedents for pronouns. This is formulated as a classification problem, with the model assigning labels to potential antecedents.

In the training phase, our model aims to minimize the cross-entropy loss, which quantifies the disparity between its predictions and the true labels. The formula for cross-entropy loss is as follows:

$$Loss = \sum_{j=1}^{C} y_j * \log p_j \tag{1}$$

where *Loss* represents the overall loss.

C is the total number of potential antecedents.

y_j is the true label for each potential antecedent, which equals 1 if it is the correct antecedent, and 0 otherwise.

p_j is the model's predicted probability for each label.

iii. **Testing**: After training, the model's effectiveness is assessed using a separate test dataset. The model is used to predict anaphor locations or labels in the test data. Model performance is evaluated using standard metrics, including accuracy, precision, recall, and F1-score. Inference: With a trained model in place, it can be applied to new text passages for anaphora resolution. The input text is preprocessed and converted into numerical representations. The model generates predictions for anaphor locations or labels. Output: The research paper presents the model's predictions and the results obtained from the algorithm, contributing to the field of anaphora resolution. This algorithm is designed to tackle anaphora resolution, offering a structured approach for researchers to implement and evaluate in their work. The research paper can provide more detailed information and findings related to the algorithm's performance and applications [13, 14].

Algorithm: CART for Anaphora Resolution

Input (dataset D): -- A dataset of text segments or sentences.

Labels: -- Labels indicating anaphoric (1) or non-anaphoric (0) references for each text segment.

max_depth: -- Maximum depth of the decision tree.

1. **Input (dataset D)**:
- Define the dataset D containing text segments and their corresponding labels.
2. **Feature Extraction**:
- Transform text data into numerical features.
3. **Data Splitting**:
- Split D into a training set D_train and a testing set D_test.
4. **Decision Tree Construction**:
- Initialize an empty decision tree T.
- Call BuildTree to construct the decision tree with parameters T, D_train, andmax_depth.
5. **BuildTree (T, D, max_depth)**:
- If max_depth is 0 or stopping criteria met (e.g., pure node), return a leaf node labeled with the majority class.
- Calculate the impurity measure (Gini impurity or entropy) for the current node D.
- If impurity is below a threshold or max_depth is reached, return a leaf node labeled with the majority class.
- Select the feature and threshold that minimize impurity (maximize information gain).

- Create a decision node in tree T with the selected feature and threshold.
- Recursively call BuildTree for left and right child nodes, passing relevant subsets of D.

6. **Stop**:
- On reaching a maximum depth or achieving a pure node.

7. **Pruning**:
- Apply pruning techniques to simplify the tree and prevent overfitting.

8. **Classification**:
- Use the trained decision tree T for classifying new text segments.
- Traverse T is based on text segment features.
- Reach a leaf node and predict the anaphoric or non-anaphoric label based on the majority class.

9. **Model Evaluation**: For this dataset = *["kahal jaahai ki Bambai sapna ke sahar hai", "je khabsurat baadi, samunder aau balibud sitara se hamesa jagmagait raha hai", "pichala saal ekar darsan kare ke mauka millai", "[get be aaf india Location NE]se bot par sabar hoke ham log [elifenta ke gufa Location_NE] dekhe lagi aage baDhli", "samundar ke bich asTimar aau badkan jahajwanke dur choD [asli jagah] lagi aau age badhi T najaara sab ke dekh ke man mein kai tarah ke tasbir aabe laglai" "[elifenta ke gufa Location_NE] kalatmak kalakriti ke srinkhala hai [je] ki [elifenta Ail and Location_NE] mein stith hai" "[ekara] [siti aaf kevas] kahal jaa hai" "mumbai ke geTabe aaf india se lagbhag 12 kilometar dur par arab sagar mein basal [e] chota Taapu hai" "[ehaan] [saat gufa] banal hai [jekara mein] se mukhya gufa mein 26 stambh hai" "[ehaan] [bhagbaan Shankar ke nau baDa-baDa murti] hai" "murti kala ke [e] nayaab namuna ke unesko bisba dharohar ke suchi mein 1987 mein saamil kail gelai" "caron or se samundar se ghiral [e Taapu] 1 kabhi [ghara puri] 1 ke naam se janal jaa halai" "[ekar] banabe ke sadike leke bises agyaban mein matbhed hai" "aaumanal jaa hai ki chaTha se leke aanThawan satabdi ke Madhya bhartiya silp kala ke namuna [ekar] gufa mein dekhal jaa saka hai" "[enka] Thos chaTatan ke taraske banabal gelai he"]*

3   Results and Discussion

In assessing the algorithm's performance, a comprehensive evaluation was conducted to measure its effectiveness in detecting non-anaphor and anaphora classes. Precision, recall, and F1-score metrics were employed for each class, alongside macro average and weighted average computations. The algorithm exhibited an overall accuracy of 92%, indicative of its robust performance across both classes. For a detailed breakdown of these metrics, please refer to Table 1.

Table 1 Result of anaphora resolution

Class	Precision	Recall	F1-score
Non-anaphor class	1.00	0.33	0.50
Anaphor class	0.92	1.00	0.96
Macro avg	0.96	0.67	0.73
Weighted avg	0.93	0.92	0.90

4 Conclusions

Magahi has a rich pronominal system, as the research demonstrates unambiguously. Not only theoretically but technically, there are still more syntactic structures that require investigation. Computational work connected to Magahi is still far behind, despite extensive research on English, European, and South Asian languages. In contrast to other Indian languages, Magahi should be considered for applications involving natural language generation, machine translation, information extraction, automated summarization, question-answering systems, and other NLP tasks that require the accurate identification and resolution of anaphoric expressions.

References

1. B. Lust, *Lexical Anaphors and Pronoun in Selected South Asian Languages: A Principled Typology*, (2000)
2. S. Chandra, Anaphora in Magahi. Ann. Arts Humanities Soc. Sci. **2**(1), 1010 (2023)
3. S. Chandra, Syntactic function of anaphora in Magahi. Annu. J. Linguist. Soc. Nepal (LSN) **30**, 31–39 (2015)
4. J. Higginbotham, Pronouns and bound variables. Linguist. Inquiry **11**, 679–708 (1980)
5. S. Lappin, H. Leass, An algorithm for pronominal anaphora resolution. Comput. Linguist. Linguist. **20**(4), 535–562 (1994)
6. T. Reinhart, *Strategies of Anaphora Resolution*, ed. by Bennis et al., *Interface Strategies*, (Amsterdam, North Holland, 2000)
7. N. Ge et al., *A Statistical Approach to Anaphora Resolution. VLC@COLING/ACL* (1998)
8. V. Ng, C. Cardie, Identifying anaphoric and non-anaphoric noun phrases to improve coreference resolution, in *International Conference on Computational Linguistics*, (2002)
9. X. Yang, Feng et al., A twin-candidate model for learning-based anaphora resolution. Comput. Linguist. **34**, 327–356 (2008)
10. X. Yang et al., Coreference resolution using competition learning approach, in *Annual Meeting of the Association for Computational Linguistics*, (2003)
11. C. Sutton et al., Dynamic conditional random fields: factorized probabilistic models for labelling and segmenting sequence data, in *Proceedings of the twenty-first international conference on Machine learning,* (2004). https://doi.org/10.1145/1015330.1015422
12. J.R. Finkel, T. Grenager, C. Manning, Incorporating non-local information into information extraction systems by gibbs sampling, in *Proceedings of the 43rd Annual Meeting of the Association for Computational Linguistics (ACL'05)*, (2005), pp. 363–370
13. L. Kumari, S. Debarma, N. Kar, R. Shyam, Machine translation evaluation system: development & comparison. Int. J. Sci. Technol. (IJSAT). **2**(1), 199–204 (2011)

14. L. Kumari, N. Sharma, Measuring the effectiveness of deep learning in STEM education: a comparative analysis of student outcomes and engagement, in STEM: A Multi-Disciplinary Approach to Integrate Pedagogies, Inculcate Innovations and Connections, 2023, pp. 163–174, https://doi.org/10.52305/UXLT2425

Green Communication: A Sustainable Approach to Environmental Awareness and Action

Akhmad Yusuf Zuhdy and Ifarrel Rachmanda Hariyanto

Abstract Green communication is a rapidly developing concept that combines communication principles with the aim of increasing awareness and pro-environmental action. Currently, global environmental challenges are increasing, including climate change, decreasing air and water quality, ecosystem damage, and loss of biodiversity. These problems encourage the need for behavioral changes and collective action to achieve environmental sustainability. Currently, existing green communication includes various initiatives, trends, and practices used by organizations, companies, governments, and individuals to integrate sustainable concepts in the field of information and communication technology (ICT). The conclusion of the research is to illustrate the importance of green communication as an approach that can influence environmental awareness and encourage pro-environmental action. In the context of increasingly pressing environmental issues, effective communication is key to influencing behavior and creating positive change in society.

Keywords Green communication · Sustainability · Global environmental challenges · Environmental awareness

1 Introduction

Green communication is a rapidly developing concept that combines communication principles with the aim of increasing awareness and pro-environmental action. Currently, global environmental challenges are increasing, including climate change, decreasing air and water quality, ecosystem damage, and loss of biodiversity. These

A. Y. Zuhdy (✉) · I. R. Hariyanto
Department of Civil Infrastructure Engineering, Sepuluh Nopember Institute of Technology, Surabaya, Indonesia
e-mail: yuf_di@yahoo.com

I. R. Hariyanto
e-mail: ifarrelrh@gmail.com

problems encourage the need for behavioral changes and collective action to achieve environmental sustainability.

One effective effort is through the implementation of green communication. This concept includes communication strategies that aim to increase public understanding of environmental issues, inspire environmentally friendly behavior, and promote sustainable policies and practices.

Currently, information and communication technology have a crucial role in disseminating information about environmental issues and encouraging public participation in sustainable action. The use of social media, websites, applications, and other digital tools has become an important means of conveying pro-environmental messages to a wider audience [1].

Therefore, this research aims to analyze the concept of green communication, understand the role of technology and media in facilitating pro-environmental communication, and identify strategies and best practices to maximize the positive impact of green communication on environmental awareness and pro-environmental actions. The role of communication in climate change interventions is shown in [2]. The focus on energy savings and sustainable practices is reported in [3]. Factors influencing pro-environmental behavior are illustrated in [4]. The importance of individual action in conservation is mentioned in [5].

2   Methodology

Green communication is a type of communication related to environmental and sustainable issues. This includes the use of communications to convey information, education, and messages that support environmentally friendly actions and contribute to sustainability. Green communication can be implemented by organizations, companies, governments, and individuals to educate, influence behavior, and promote more sustainable practices in society. The aim is to raise awareness about environmental issues and encourage positive changes in behavior and actions that support a healthier and more sustainable environment.

This research will use qualitative and quantitative approaches to analyze the concept of green communication, the role of technology, best practices, impact evaluation, and the challenges faced. Interview, survey, observation, and content analysis methods will be used to understand this phenomenon in more depth.

• Data Sources

The research utilizes both primary and secondary data sources. Primary data comprises interviews with environmental communication experts and green communication practitioners, community surveys assessing pro-environmental awareness and participation, and direct observations of green communication practices in organizations and society. Secondary data consists of literature studies on green communication concepts and related research, along with analysis of documents and online content concerning green communication strategies and their implementation.

- Data Collection Technique

This research utilized several data collection methods, including structured interviews with green communication experts, complemented by audio/video recordings and transcriptions for analysis. A questionnaire survey was conducted to gather data on environmental awareness and behavior, distributed through both online and offline platforms. Additionally, direct observation of green communication practices within organizations and communities was employed, with field notes and analyses of communication strategies and effectiveness.

- Data Analysis

The data analysis used in this research is divided into qualitative analysis, which includes document content analysis, interviews, and observations to gain in-depth insight into the green communication concept and its implementation strategies. Furthermore, quantitative analysis includes processing survey data using statistical methods to get a more quantitative picture of the level of awareness and pro-environmental behavior.

- Validity and Reliability

Validity and reliability are used to guarantee the data that has been collected, where the validity of the data will be guaranteed through triangulation, namely the use of several methods and data sources. Reliability will be guaranteed through consistency in data collection, analysis, and interpretation.

This research methodology includes the type of research, data sources, data collection techniques, and analysis that will be used to gain a comprehensive understanding of green communication and how it can influence environmental awareness and pro-environmental actions.

3 Results and Discussion

From the results of observations using the various methods mentioned, analysis was carried out on the results of the questionnaire, which consisted of respondent information, environmental awareness, pro-environmental attitudes, use of media in green communication, and green communication itself.

- Part A: Respondent Information

The results of the questionnaire showed that most respondents were 18–25 years old, with a male gender, and their highest education was senior high school.

- Part B: Environmental Awareness

The questionnaire results indicate a high level of environmental awareness among respondents, but over 50% reported insufficient recycling of household waste, infrequent use of eco-friendly vehicles, and limited support for forest and ecosystem protection campaigns. Despite this, over 50% expressed significant concern about climate change, frequently reduced single-use plastic consumption, and understood the environmental impact of plastic waste. Moreover, a majority reported regular reductions in water and electricity consumption, active involvement in environmental activities, and support for renewable energy. However, more than 50% felt the government's role in environmental protection was inadequate.

- Part C: Pro-Environmental Attitude

The results of the questionnaire show that more than 50% of respondents are quite concerned about environmental problems, are quite ready to change their behavior to support the environment, and support using renewable energy. Apart from that, respondents also indicated that they were quite accustomed to reducing waste and reusing, prioritizing environmentally friendly products when shopping.

- Part D: Use of Media in Green Communication

The questionnaire reveals that while respondents frequently consume environmental news through media and follow related social media and websites, their participation in online conservation campaigns is lower. Over half regularly verify the credibility of their environmental information sources and often share such content on social media. They favor environmentally responsible brands, believe the media sufficiently covers environmental issues, and prefer "environmentally friendly" products. Additionally, respondents recognize media's influence on their environmental attitudes and express a desire for more environmental programming.

- Part E: Green Communication

The survey in section E indicates that over half of the participants prefer electronic over paper communication, endorse email to cut down on paper use, and frequently choose webinars or online meetings over in-person ones. They commonly utilize apps to reduce paper consumption and favor telecom providers with green policies to lessen their carbon footprint. Respondents also back efforts to decrease paper advertising and strongly favor environmentally friendly devices, supporting energy conservation in communications and viewing digital methods as more eco-friendly than traditional ones.

A.   The Role of Media and Technology in Green Communication

Role of Media in Green Communication:

1.   Liaison between stakeholders:

The media functions as a link between organizations, government, scientists, activists, and society. Information related to environmental issues and solutions can be accessed by the public through the media, allowing them to get involved and contribute.

2. Increasing Environmental Awareness:

The media plays an important role in raising public awareness about environmental issues, including the impacts of climate change, the water crisis, and biodiversity. News, articles, and special programs on the environment help in disseminating correct and relevant information.

3. Environmental Advocacy and Campaigns:

The media is used to advocate for environmental issues through public campaigns. Environmental organizations and groups use the media to mobilize support, call for action, and influence environmental policy.

4. Promotion of Eco-friendly Technology:

The media facilitates promotion and information about environmentally friendly technologies, such as renewable energy, electric cars, and other green innovations. This helps in forming a positive outlook and a more sustainable adoption of technology.

Role of Technology in Green Communication:

1. Global Connector:

Technology, especially the Internet and social media, enables the rapid exchange of information and communication throughout the world. It allows people to connect, share knowledge, and engage in discussions regarding environmental issues globally.

2. Effective Information Broadcasting:

Technology enables more effective broadcasting of information about sustainability initiatives, research, and environmental action through online platforms, videos, and dedicated websites. This helps to reach a wider and deeper audience.

3. Environmental Monitoring and Measurement:

Technology is used for monitoring and measuring environmental conditions such as air quality, water quality, and sustainability. Sensors and related devices provide accurate data to support pro-environmental policies and actions.

4. Reduction of Carbon Footprint:

Technology is used to develop solutions that help reduce carbon footprints, such as applications for energy management, application-based transportation systems, and energy efficiency technologies.

Conclusion:

Media and technology play an important role in green communication. They serve as a means of disseminating information, building awareness, and facilitating pro-environmental action. With proper use, media and technology can play a key role in shaping more sustainable attitudes and behaviors in society.

B. Environmental Education and Awareness through Green Communication

Environmental education and awareness through green communication are at the core of triggering positive and sustainable action on environmental issues. The following is a further explanation regarding this:

Environmental Education through Green Communication:

1. In-depth information Dissemination:

Green communication functions as a tool to convey in-depth information about environmental issues, including causes, impacts, and existing solutions. Clear and reliable information forms the basis of a solid education.

2. Teaching Environmental Concepts:

Green communication helps in teaching critical concepts such as biodiversity, material circulation, and the impact of climate change. This increases people's understanding of the complexity of ecosystems and environmental challenges.

3. Action Education and Behavior Change:

Green communication not only provides knowledge but also encourages action. This education includes more environmentally friendly practices, such as recycling, saving energy, using public transportation, and reducing waste.

4. Green Literacy Development:

Green communication helps develop green literacy in the community. This literacy includes an understanding of sustainability, environmental ethics, and the impact of everyday consumption decisions.

Environmental Awareness through Green Communication:

1. Increased Awareness of Environmental Impact:

Green communication helps increase public awareness about the impacts produced by human activities on the environment. This includes increasing awareness about land degradation, climate change, air and water pollution, and biodiversity loss.

2. Understanding Global Interconnections:

Through green communication, society can understand the global interconnectedness of environmental issues. They recognize that local actions can have a global impact and that global collaboration is needed to solve environmental challenges.

3. Be Aware of the Urgency of Action:

Green communication highlights the urgency of action on environmental issues. Society is becoming more aware of the urgent need to take sustainable and proactive action to protect and restore the environment.

4. Provide Inspiration and Positive Examples:

Through success stories, innovative projects, and successful campaigns, green communication provides inspiration and positive examples to the community. This encourages individuals and organizations to adopt more sustainable actions.

C. Implementation of Green Communication

Green communication focuses on prioritizing environmental sustainability, characterized by environmental awareness, transparency, educational initiatives, community involvement, consistency in sustainable practices, avoiding greenwashing, and collaborative efforts. It aims to promote environmental responsibility and sustainability. Surabaya's achievement of winning the ASEAN Environmentally Sustainable City award in 2021 as the cleanest city in Southeast Asia exemplifies the successful application of green communication. The city's strategies included increasing public awareness, encouraging public transportation use to reduce emissions, and offering vehicle emission education, demonstrating how green communication principles can be effectively implemented to enhance environmental sustainability.

4   Conclusion

This research illustrates the importance of green communication as an approach that can influence environmental awareness and encourage pro-environmental actions. In the context of increasingly pressing environmental issues, effective communication is key to influencing behavior and creating positive change in society.

Suggestions and recommendations:

a. The Importance of Green Communication:

Green communication plays a crucial role in forming environmental awareness in society. By conveying appropriate, clear, and targeted messages, information about environmental issues and solutions can be effectively conveyed to various levels of society.

b. Encouraging Pro-Environmental Action:

Green communication has great potential for encouraging environmentally friendly actions. Messages that inspire and guide people to change behavior and take sustainable action are very necessary to maintain environmental balance.

c. Technology as a Key Driver:

The role of technology, especially social media and digital platforms, is very significant in expanding the reach of green communication. Technology enables faster message distribution, greater interaction, and broader community engagement.

d. Challenges and Opportunities:

Challenges such as information overload, message credibility, complex language, and limited technological accessibility are obstacles that must be overcome. However, with the right strategy and the implementation of appropriate recommendations, these challenges can be turned into opportunities to increase the effectiveness of green communication.

e. Important Role of Education and Awareness:

Environmental education and awareness are the foundations for effective green communication. Communities need to be empowered with accurate knowledge about environmental issues as well as awareness of the actions they can take to overcome environmental challenges.

Implications and Recommendations:

Based on this conclusion, it is important to continue to develop and improve green communication strategies that can more effectively influence public awareness and move them to contribute to environmental sustainability. Organizations, governments, and individuals need to come together to ensure that pro-environment messages are spread widely, and that real action is taken to preserve and protect the environment for a better future.

Conclusion and Call to Action for the Future:

In undertaking the journey through the green communication concept, we have gained deep insight into the crucial role of communication in bridging environmental awareness and real action. Environmental issues are currently increasingly pressing, and green communication is a powerful tool to mobilize society to play an active role in protecting the planet.

It is important for us to understand that every individual has a role in creating positive change. Through greater awareness, collaboration, and targeted action, we can bring about profound and lasting change. We invite you to stand together and become agents of change in protecting the environment and passing it on to future generations.

Call to Action for the Future:

1. Educate and Counsel: Engage your network in learning about environmental issues to enhance awareness.
2. Minimize Environmental Impact: Adopt simple practices like reducing plastic use and saving resources to make a collective difference.
3. Support Environmental Efforts: Participate in campaigns and back organizations advocating for the environment to influence change.
4. Collaborate for Impact: Work together with various stakeholders for innovative environmental solutions.
5. Inspire Through Stories: Share personal sustainability efforts to motivate others.
6. Utilize Technology: Use digital tools to promote environmental consciousness and action.

References

1. B. Kashif, S.U. Khan, S.A. Madani, K. Hayat, M.I. Khan, N. Minallah, J. Kolodziej, L. Wang, Z. Sherali, D. Chen, A survey on green communications using adaptive link rate. Clust. Comput. **16**, 575–589 (2013). https://doi.org/10.1007/s1058601202258
2. E. Maibach, C. Roser-Renouf, A. Leiserowitz, Communication and marketing as climate change–intervention assets: a public health perspective. Am. J. Prev. Med. **35**(5), 488–500 (2008). https://doi.org/10.1016/j.amepre.2008.08.016
3. D.L. Gadenne, B. Sharma, D. Kerr, T. Smith, The influence of consumers' environmental beliefs and attitudes on energy saving behaviours. Energy Policy **39**(12), 7684–7694 (2011). https://doi.org/10.1016/j.enpol.2011.09.002
4. D. Arli, F. Tjiptono, A.V. Hubeis, Understanding pro-environmental behavior: a cross-cultural analysis between Indonesia and Australia. Int. J. Consum. Stud.Consum. Stud. **43**(4), 411–419 (2019). https://doi.org/10.1108/09590551211222367
5. P.W. Schultz, Conservation means behavior. Conserv. Biol. **25**(6), 1080–1083 (2011). https://doi.org/10.1111/j.1523-1739.2011.01766.x

Digital Amplification of Silent Struggles

Arundhati Sharma, Roli Mishra, Rahul Sharma, Anchit Pandey, and Upanshu Mishra

Abstract The study shows the impact of digital media in giving voices to the silenced marginalized victims of violence. Netflix's "13 Reasons Why" has ignited crucial discussions, primarily around adolescent suicide, yet it often overlooks the prevalent theme of trauma, especially related to sexual violence. This study applies a feminist perspective to analyze how the series portrays trauma and its roots in toxic masculinity, which contributes to the devaluation of female bodies and sexual autonomy. The narrative, particularly through the character of Hannah Baker, reflects the pervasive impact of societal issues like slut-shaming and institutional neglect. However, the emergence of the #MeToo movement has been instrumental in providing a platform for healing and empowerment for the victims of such trauma. This movement, in the digital era, offers a space for survivors to voice their experiences and initiate a healing process. It stands as a counterforce to the silencing effect of toxic masculinity and societal indifference. Thus, "13 Reasons Why" is more than a series; it's a reflection of the often-unspoken traumas of sexual violence, exposing the invisible influence of toxic masculinity and societal apathy, while also emphasizing the #MeToo movement's role in fostering healing and empowerment.

A. Sharma (✉) · R. Mishra · R. Sharma · U. Mishra
Amity University Patna, Patna 801503, Bihar, India
e-mail: arundhatisharma2092@gmail.com

R. Mishra
e-mail: rmishra@ptn.amity.edu

R. Sharma
e-mail: rahulsharma.kv1987@gmail.com

U. Mishra
e-mail: umishra@ptn.amity.edu

A. Sharma
SRM University, Sonepat 131029, India

A. Pandey
Patna University, Patna, Bihar 800004, India
e-mail: anchitpandey@pup.ac.in

© The Author(s), under exclusive license to Springer Nature Switzerland AG 2025
S. Sambhav et al. (eds.), *Empowering Solutions for Sustainable Future in Science and Technology*, SpringerBriefs in Applied Sciences and Technology,
https://doi.org/10.1007/978-3-031-77837-7_12

Keywords Toxic masculinity · Sexual violence · Slut-shaming · Trauma ·
#MeToo movement · Digital media

1 Introduction

In a period where breaking silences is encouraged, it remains controversial to discuss
certain topics openly, such as acts of violence committed by women. The #MeToo
movement, which gained prominence in 2017 following Alyssa Milano's tweet, has
been pivotal in raising awareness about sexual harassment and abuse, particularly in
the workplace (Fig. 1). It led to significant cultural and workplace changes, shedding
light on the widespread issue of sexual violence.

#MeToo—this digital expansion has brought new dimensions to the understanding
of violence, inclusivity, and the human experience in modern society. Individuals
who do not actively work to disengage from all forms of oppression cannot be fully
committed to eradicating violence.

Pierre Bourdieu (1998) and his associates critically analyzed the social and political impacts of media, focusing on its "symbolic power" as noted by Nick Couldry
in 2003. They viewed the media as a cultural production sphere, consisting either of
a single field or multiple fields, each with its own prestige, status, and values. This
"symbolic power" is seen as a self-fulfilling prophecy, reliant on public complicity
and legitimation, as described by Nick Crossley in 2001.

In analyzing the media's portrayal of trolling, we adopt the concept of "symbolic
violence" from Bourdieu's framework. This term is specifically used to describe
how such violence is directed at women and minority groups in media narratives
about trolling, leading to their devaluation and stigmatization. Media sometimes
ignores Black voices in #MeToo; Burke and Haynes highlight bias (Fig. 2). Gender
stereotypes reinforce traditional roles, shaping media bias on abuse and marginalizing

Fig. 1 Tweet of Alyssa
Milano that initiated the
digital movement #Metoo

Alyssa Milano ✓
@Alyssa_Milano

Follow

If you've been sexually harassed or
assaulted write 'me too' as a reply to
this tweet.

Me too.

Suggested by a friend: "If all the women who
have been sexually harassed or assaulted
wrote 'Me too.' as a status, we might give
people a sense of the magnitude of the
problem."

1:21 PM · 15 Oct 2017

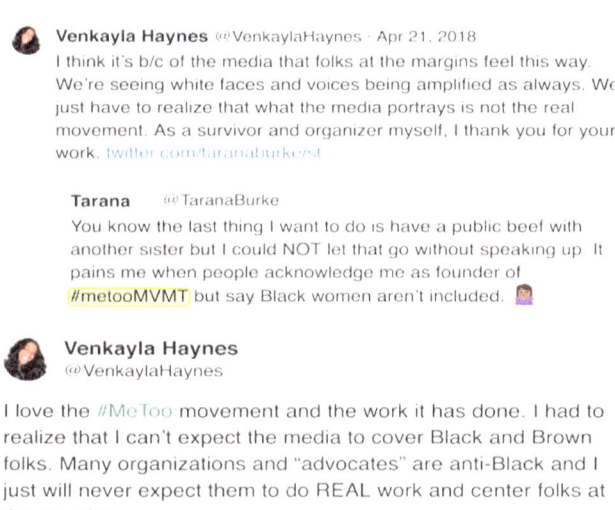

Venkayla Haynes @VenkaylaHaynes · Apr 21, 2018
I think it's b/c of the media that folks at the margins feel this way.
We're seeing white faces and voices being amplified as always. We
just have to realize that what the media portrays is not the real
movement. As a survivor and organizer myself, I thank you for your
work. twitter.com/taranaburke/st

Tarana @TaranaBurke
You know the last thing I want to do is have a public beef with
another sister but I could NOT let that go without speaking up. It
pains me when people acknowledge me as founder of
#metooMVMT but say Black women aren't included. 🙍🏽

Venkayla Haynes
@VenkaylaHaynes

I love the #MeToo movement and the work it has done. I had to
realize that I can't expect the media to cover Black and Brown
folks. Many organizations and "advocates" are anti-Black and I
just will never expect them to do REAL work and center folks at
the margins.

7   1:28 PM · Apr 21, 2018

See Venkayla Haynes's other Tweets

Survivor and organizer Venkayla Haynes comments on the #MeToo movement

Fig. 2 The effect of the movement on women of color and the role of digital media

Black voices, as seen in Venkayla Haynes' critique and Tarana Burke's frustration with #MeToo's exclusion of Black women.

The show underscores the compelling need for an intensified focus on addressing gender-based violence and fostering an environment of awareness, support, and preventative action in our schools and communities to safeguard the mental health of young individuals impacted by such experiences. In this concise critique, we focus on what we identify as the triad of primary themes embedded in "13 Reasons Why". These themes encompass (1) the pervasive influence of toxic masculinity, (2) the practice of slut-shaming as a mechanism to demean the female form and female sexual autonomy, and (3) MeToo as a means to heal from trauma. While much of the existing discourse surrounding the series is anchored in its portrayal of suicide [1], we contend that an in-depth exploration of gender-based violence and aggression is warranted and essential to decoding the underlying motivations and actions of the central character, Hannah Baker. This analysis is instrumental in unveiling the nuanced layers of trauma that stem from systemic failures and gendered violence, further amplifying the protagonists mental health decline.

2   Methodology

In exploring the intersection of slut-shaming, toxic masculinity, digital media, and the #MeToo movement as a methodology, we can understand the complex dynamics of gender-based violence and societal attitudes. Slut-shaming, often perpetuated by toxic masculinity, involves stigmatizing individuals, especially women, for their perceived sexual behavior. This phenomenon is amplified by digital media, which provides a platform for widespread dissemination and often exacerbates these issues. However, the #MeToo movement, greatly enabled by digital media, offers a counter-narrative and a space for challenging these harmful norms. By combining these elements, we can critically analyze how societal attitudes shape gender relations and the potential of digital platforms for both harm and empowerment. Our analysis underscores the persistent narrative of toxic masculinity throughout the Netflix adaptation, shedding light on the detrimental effects of masculine norms depicted and their role in shaping the mental well-being of the young female protagonist, Hannah Baker. Hannah's ordeal is not rooted in a solitary incident but rather a series of sexual violence experiences fueled by microaggressions, profoundly impacting her psychological well-being.

An illustrative instance from the first episode features Justin, portrayed as Hannah's initial romantic interest, capturing a seemingly innocent photograph of Hannah sliding toward him. The unintended visual implication of the image, showing her skirt uplifted, fuels a narrative of sexual conquest. Entrapped by the unyielding expectations of toxic masculinity, Justin chooses the affirmation of his male peers over Hannah's dignity, letting the damaging rumor fester. This relentless exposure to toxic masculinity and the associated sexual violence instigates a cascade of trauma for Hannah, exacerbating her mental health decline [2].

In the third episode, Alex, shown as a good friend to Hannah, creates a "hot-or-not" list, ranking girls based on their physical attributes. This list, intended to boost his standing among his male peers and to retaliate against his ex-girlfriend, Jessica, for not sleeping with him, marks Hannah as having the "best ass". The impact is profound; Hannah is deeply distressed by her objectification, ending her ties with Alex.

In Episode 9, Hannah is depicted seeking refuge in Jessica's room during a party. However, her sanctuary becomes a horrifying spectacle as she becomes a silent witness to sexual assault. Jessica, too intoxicated to consent, is vulnerable when her boyfriend, Justin, leaves the room. Bryce, a character embodying affluent and aggressive masculinity, takes advantage of Jessica's incapacitated state. Despite Justin's initial resistance, he is subdued by Bryce's invocation of the "bro code" and the toxic norms of male entitlement. This grim depiction underscores the perverse consequences of toxic masculinity—a world where male entitlement overrides consent, and the defense of masculine norms inhibits proactive intervention against sexual violence [3]. The portrayal of Jessica's assault is not just a visual recounting of an isolated incident but is emblematic of the overarching culture of silence and complicity surrounding sexual violence.

Episode 12, intensifying the narrative of sexual violence, captures Hannah's rape by Bryce. The portrayal of Hannah's dissociation during the assault is a harrowing encapsulation of the trauma victims endure [4]. Here again, toxic masculinity is laid bare—Bryce's unabashed entitlement and denial underscore the sinister norms that foster sexual violence. When Clay confronts Bryce, another dimension of masculinity is unveiled—one that seeks to defend women's honor yet is equally imprisoned within its constraining norms. Bryce's blatant denial and dismissal of his heinous act as rape reflect the deep-rooted societal norms perpetuating sexual violence, framing it as an entitlement rather than a crime. "13 Reasons Why" serves as both a reflection and an indictment of a society where toxic masculinity, entitlement, and silence intertwine, necessitating movements like #MeToo to unveil the harrowing truths, break the silence, and embark on a collective journey of acknowledgment, healing, and transformation.

The series "13 Reasons Why" lays bare the devastating impact of toxic masculinity, vividly illustrating its connection to violence and mental health struggles. Throughout the episodes, viewers are confronted with the damaging effects of oppressive gender norms, manifesting in various forms of harm and psychological turmoil. Despite the stark portrayal of toxic masculinity, there is an ironic absence of similar outrage or censorship that other controversial content might provoke. Nevertheless, despite their documented associations with adverse mental health outcomes, the insidious norms of toxic masculinity, including potential links to suicide, remain unchallenged and pervasive in mainstream media.

Slut-shaming, the persistent cultural presence of the sexual double standard, where differing sexual norms and expectations are applied to men and women, has been a subject of interest for researchers for years [5]. This double standard is confined to adult society and extends into adolescent social dynamics, impacting peer acceptance and popularity. The media, too, plays a role in reinforcing this dynamic, often portraying women within the confines of a virgin–whore dichotomy, thereby implying their complicity in their victimization.

The series also portrays instances where young women are set against one another, compelled to defend their sexual reputations. A striking example is the aftermath of the disclosure of the "best/worst" list, prompting Jessica to confront Hannah with accusations of infidelity with her boyfriend, Alex. Rather than holding Alex accountable for the list, Jessica severs her friendship with Hannah. This action punishes Hannah for an assumed sexual misdemeanor and deprives her of essential social support. In the swirl of unsubstantiated rumors regarding her supposed sexual indiscretions, Episode 6 depicts Hannah falling victim to sexual assault during her initial date with Marcus, a popular student. Eager to corroborate the circulating gossip portraying Hannah as "easy," Marcus inappropriately touches her in public. His astonishment at Hannah's vehement reaction underscores a disturbing disconnect and entitlement.

The intersection between toxic masculinity, consent, and the mental health of young women is portrayed. Throughout the series, Hannah's sexuality and desires are manipulated and weaponized against her. The cumulative effect of these traumatic experiences is intricately linked to her ultimate decision to commit suicide.

While discussions around the series have touched upon bullying, it is crucial to delve deeper and categorically identify this as sexual harassment and violence. Acknowledging and naming the specific nature of this abuse is a pivotal step toward fostering dialogues that can catalyze meaningful change and offer support to those grappling with similar experiences, aligning with the narratives and objectives of the #MeToo movement.

3   Conclusion

In conclusion, Netflix's "13 Reasons Why" serves as a significant cultural artifact that extends beyond its depiction of adolescent suicide to address the pervasive issue of trauma, particularly rooted in sexual violence. Through a feminist lens, the series highlights the detrimental effects of toxic masculinity and societal practices like slut-shaming, which undermine female agency and perpetuate trauma. The #MeToo movement's emergence, amplified by digital media, offers a crucial platform for survivors to express their experiences and begin healing. This movement challenges the silence imposed by toxic masculinity and societal indifference, thus underscoring the series' role in reflecting and responding to the unspoken traumas associated with sexual violence. "13 Reasons Why" transcends being just a series; it symbolizes the struggle for acknowledgment and healing in the face of sexual violence, further empowered by the transformative influence of the #MeToo movement.

The #MeToo movement can aid in healing by providing a platform for survivors of sexual violence to share their experiences, thereby breaking the silence and stigma surrounding such incidents. This collective sharing can foster a sense of solidarity and validation, as survivors realize they are not alone. Including #MeToo narratives becomes particularly salient when examining the experiences of female characters like Jessica and Hannah.

The #MeToo movement, in this context, emerges as a pivotal platform, not only for acknowledgment but for collective healing. It transcends individual narratives, weaving a tapestry of collective experience, silent yet resonant [6]. In the intricate dance of trauma and memory, characters like Jessica and Hannah become reflective surfaces, each echoing the silent yet potent narratives of sexual assault and harassment that define the contemporary social landscape. "13 Reasons Why" offers not just a portrayal of individual traumas but an exploration of the systemic structures and norms that define and shape these experiences. Jessica and Hannah's narratives are not isolated—they are interwoven into the broader societal narrative, echoing the silent yet potent dialogues of the #MeToo movement. Each character, each echo of trauma, is a testament to the pervasive norms that silence and define an invitation for reflection, confrontation, and transformation [7].

References

1. J.W. Ayers, B.M. Althouse, E.C. Leas, M. Dredze, J.-P. Allem, Internet searches for suicide following the release of 13 reasons why. JAMA Intern. Med. **177**(10), 1527 (2017). https://doi.org/10.1001/jamainternmed.2017.3333

2. G. Chandra, I. Erlingsdóttir, *The routledge handbook of the politics of the #MeToo movement*, (2020). https://doi.org/10.4324/9780367809263

3. M. Rodino-Colocino, Me too, #MeToo: countering cruelty with empathy. Commun. Crit.Al/Cult. Stud. **15**(1), 96–100 (2018). https://doi.org/10.1080/14791420.2018.1435083

4. N. Tec, *Recapturing the past,* (Oxford University Press eBooks, 2009), pp. 27–48. https://doi.org/10.1093/acprof:oso/9780195389159.003.0003

5. M. Randall, Sexual assault law, credibility, and ideal victims: consent, resistance, and victim blaming. Can. J. Women Law **22**(2), 397–433 (2010). https://doi.org/10.3138/cjwl.22.2.397

6. M. Crawford, D. Popp, Sexual double standards: a review and methodological critique of two decades of research. J. Sex Res. **40**(1), 13–26 (2003). https://doi.org/10.1080/0022449030955 2163

7. M.A. Nash, Gender trouble: feminism and the subversion of identity, ed. by J. Butler (Routledge, New York, 1990)—Homophobia: a weapon of sexism, ed. by S. Pharr, C.A. Inverness (Chardon Press, 1988) Hypatia: A J. Fem. Philos. **5**(3), 171–175 (1990). https://doi.org/10.1017/s08875 36700007339

Prediction of Air Quality Index Using Support Vector Machines

Saurabh Sambhav, Shilpi Singh, Shashi Bhushan, and Santosh Dixit

Abstract This study proposes a system for the analysis and prediction of the air quality index (AQI) throughout the cities of India. It takes the AQI of different cities in India as an input dataset and analyses the data to understand the patterns to predict the quality of air. Atmospheric pollution contributes to a variety of respiratory issues, including heart disease and lung cancer. Therefore, finding a technique to monitor the air quality index is of the highest importance because it is always preferable to be aware of the degree of pollution as soon as possible so that preventative steps can be implemented. In India, we suggest an analysis and prediction framework for the air quality index (AQI) utilizing support vector machines (SVM) because of their capability to handle high-dimensional data and nonlinear interactions. Overall, this study demonstrates how our machine learning algorithms can be utilized as a tool for precision monitoring and for detecting air pollution index. Early detection of air quality conditions enables us to take necessary action to stop the disease spread and reduce air pollution illness.

Keywords Air pollution · AQI · Support vector machines · Air contaminants · ML

S. Sambhav (✉) · S. Singh · S. Bhushan · S. Dixit
Amity School of Engineering and Technology, Amity University, Patna, Bihar 801503, India
e-mail: ssambhav@ptn.amity.edu

S. Singh
e-mail: ssingh3@ptn.amity.edu

S. Bhushan
e-mail: sbhushan@ptn.amity.edu

S. Dixit
e-mail: skdixit@ptn.amity.edu

© The Author(s), under exclusive license to Springer Nature Switzerland AG 2025
S. Sambhav et al. (eds.), *Empowering Solutions for Sustainable Future in Science and Technology*, SpringerBriefs in Applied Sciences and Technology,
https://doi.org/10.1007/978-3-031-77837-7_13

107

1   Introduction

The global concern over air pollution is growing, and India is particularly impacted by high levels of air pollution. Urban pollution of the air has been found to have a direct influence on human health, particularly in developing and industrialized areas where norms for air quality are unavailable or only partially implemented or enforced [1]. Considering the consequences of air pollution, the WHO established criteria for air quality [2]. According to the laws and rules established at all levels, including the global, national, and local, it is monitored and measured. The air quality index (AQI) is a metric for the measurement of local air quality that is calculated by the concentrations of different contaminants in the air. It is important for human health and environmental preservation. It indicates the amount of health risk linked with air pollution exposure [3].

The government uses the AQI to inform the public about the state of the air. It deteriorates with an increase in the concentration of pollutants. It represents the severity of pollution for ordinary people. It considers numerous air contaminants, including particulate matter, sulphur dioxide, ozone, carbon monoxide, etc., and provides a value to indicate overall air quality [4].

The quality of air in India varies greatly across the country and even within cities. Some locations have reasonably pure air, while others are very polluted, especially during winter. Industrial emissions, automotive traffic, construction activities, and open burning of garbage and biomass are the primary contributors to air pollution in India. It is crucial to monitor the local AQI and take the appropriate actions to protect people from the hazardous effects of air pollution. Using air purifiers, wearing masks while going outside, limiting outdoor activity during peak pollution hours, and decreasing personal car use are some frequent strategies.

The AQI in India is derived using the concentrations of eight contaminants. Particulate matter (PM) 10, (PM) 2.5, O_3, NO_2, CO, and NH_3 are the key pollutants influencing an area's AQI [5]. It is crucial to remember that AQI readings vary by nation and area, based on the unique air quality regulations and monitoring techniques utilized. It is always a good idea to check the AQI in a region before indulging in outdoor activities, especially if one is predisposed to respiratory or other health problems. Due to the growing concern about pollution and its negative effects on the health of living beings and the environment, AQI prediction using machine learning (ML) is a major topic of research. The AQI is a measure of pollution in the air that indicates how clean or dirty the air in a certain location is. ML models are used to predict AQI values by analysing historical data on various air pollution contaminants. Most linear algorithms used in the monitoring of air temperature yield significantly varying results based on the parameters [6]. The main goal of this project is to develop a prediction model that can effectively estimate air pollution levels for a specific region by analysing a variety of input variables, including geographic features, pollutant emissions, and previous air quality measurements. The data that is being used in the project is collected from various sources. Real-time monitoring of air quality by the appropriate national agencies is necessary to allow for the timely implementation of appropriate interventions to lessen harm to both humans and the ecosystem [7].

2 Literature Survey

Among the most current research on this subject are:

- The article by C. Li, Y. Li, and Y. Bao titled "Research on Air Quality Prediction Based on Machine Learning" emphasises the use of machine learning approaches for forecasting air quality. In this research paper, the authors examine different machine learning techniques and how efficiently they predicted air quality [8]. The study described in the article aims to raise the precision of air quality forecasting algorithms.
- According to Bhalgat et al. (2019), machine learning techniques were used to forecast the concentration of SO_2 in the environment of Maharashtra, India, and it was discovered that several cities in the region were very polluted and needed to be addressed [9]. However, the model they utilized did not provide the predicted results.
- Soundari et al. (2019) created a model based on neural networks (NNs) to forecast India's air quality index (AQI). When previous data on pollutant concentrations were available, they claimed that their suggested model could forecast the AQI for the entire country, any province, or any geographical location [10].
- Mahalingam et al. (2019) built and tested a model to predict the AQI of smart cities using a medium Gaussian support vector machine (SVM) in Delhi, India [11]. The scientists stated that their model was highly accurate and may be utilized in other smart cities.
- The research entitled "Air Quality Monitoring and Prediction Using SVM" by M. Kulkarni, S. Pawar, N. Rajule, S. Chavan, and A. Raut presented the machine learning technology, a support vector machine to monitor and predict the air quality [12]. The objective of the study is to create a SVM prediction model that can predict air quality levels with accuracy.

AQI directly affects the human body. The AQI assigns a numerical number to the degree of contaminants present in the air and their potential effects on human health [13]. Table 1 displays the AQI ratings for each contaminant as well as their ambient concentrations. There are various ways of forecasting AQI, support vector machines being one of the most common. Using the power of SVM to learn from previous measurements and patterns, the system can generate accurate outputs [14]. This may enhance public pressure for pollution reduction and air quality improvements [10].

Table 1 Indian AQI range
and associated outcomes

Range	Category
0–50	Good
51–100	Satisfactory
101–200	Moderately polluted
201–300	Poor
301–400	Very poor
401–500	Severe

3   Methodology

To effectively estimate air quality levels, the methodology for an air quality index prediction system employing support vector machines (SVM) includes data analysis, machine learning methods, and predictive modelling. This methodology attempts to give accurate and timely predictions of air quality, helping the government to make educated decisions and take preventative measures to decrease pollution concerns. It is done by using the capabilities of SVM, a powerful and versatile machine learning algorithm.

The methodology for this project is as follows:

a)   Data Collection

Data collection for an AQI prediction model requires finding data sources, gathering information on pollutant concentrations, and other pertinent variables, guaranteeing data quality, and taking ethical considerations into account. Building an accurate and trustworthy AQI prediction model depends on gathering relevant and high-quality data.

b)   Dataset Description

The pollutant data: data information gathered for training the system to identify air quality. NO_2, NO, NH_3, CO, SO_2, NO_x, benzene, toluene, xylene and O_3 were to be included in the data collection.

The dataset from several locations and the data present in the official government sites have been acquired. Then average readings of current air quality indices such as SO_2, particulate matter 2.5, and particulate matter 10 have been obtained. Because of defective detectors, numerous data points in the dataset are either empty or NAN. To get rid of the outliers, we must preprocess the data.

c)   Data Preprocessing

The two most frequent mistakes in data extraction and monitoring applications are outliers and missing data. The data preprocessing procedure involves a number of steps on the data, including filling in missing data (NAN) and modifying or eliminating outlier data.

d) Data Cleaning

Finding and fixing inaccurate or missing data is known as data cleaning. This includes addressing inconsistencies in data, eliminating duplicates, and imputing missing values. As data is gathered from numerous sources, data integration is necessary because the data must be aligned based on shared factors like region or time.

e) Feature Engineering

The process of incorporating or altering current features to create new features is known as feature engineering. For example, the difference between the minimum and highest temperature can be determined as a temperature range feature or hourly levels of pollutants can be averaged throughout a day to get daily averages.

f) Data Splitting

To assess the effectiveness of the machine learning model and prevent overfitting, data splitting is required. It is crucial to convert category variables into numerical values since machine learning algorithms must have numerical data. Finally, to simplify the model and reduce the number of features, dimensionality reduction techniques were applied. This made the model easier to understand and use in real-world scenarios.

g) Data Visualization

Data visualization is the process of displaying data graphically or visually using objects like maps, charts, and graphs. As a result, patterns or trends that weren't readily evident from the raw data were highlighted and the data became more accessible and understandable. The city's air quality has been assessed as shown by the air quality index for Delhi from Figs. 1, 2, 3, 4 and 5.

Fig. 1 Graph showing city's air quality where the x-axis is labelled "AQI," while the y-axis is labelled "PM2.5"

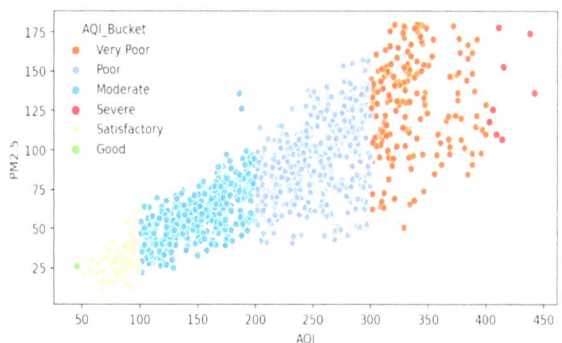

Fig. 2 Graph showing city's air quality where the x-axis is labelled "AQI," while the y-axis is labelled "NO$_2$"

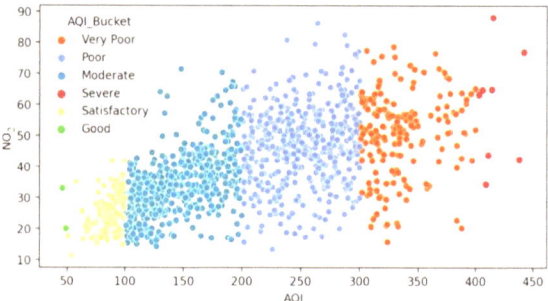

Fig. 3 Graph showing city's air quality where the x-axis is labelled "AQI," while the y-axis is labelled "NH$_3$"

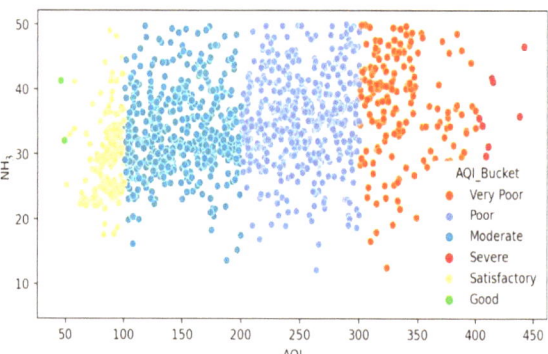

Fig. 4 Graph showing city's air quality where the x-axis is labelled "AQI," while the y-axis is labelled "NO$_x$"

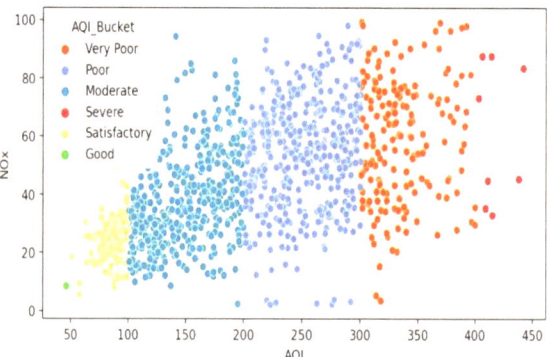

h) Machine Learning Model

The ML model was developed to forecast the AQI after the data has been pre-processed.

i) Support Vector Machine

A supervised learning method known as SVM can be applied to classification or regression problems. In order to use SVM to forecast AQI, firstly data was gathered

Fig. 5 Graph showing city's air quality where the x-axis is labelled "AQI," while the y-axis is labelled "PM 10"

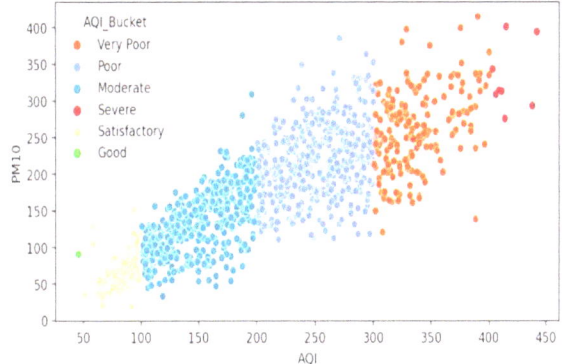

on various air contaminants such as sulphur dioxide, nitrogen dioxide, ozone, and particulate matter.

j) Support Vector Regression

In support vector regression (SVR), the objective is to identify a function that closely resembles the fundamental relationship between the input parameters and the AQI outcome. The hyperparameter C determines how wide the margin is around the hyperplane.

4 Results and Discussion

In this research, we were concerned about the model's accuracy. We explored several different approaches to discover the most accurate one so that it would be more dependable for the real-time project. The project's graphs hold the key to obtaining the finest outcomes. We plotted a graph with the dataset that was provided and an AQI versus contaminants present in the air that are responsible for the pollution, then used the Pearson correlation coefficient approach to decrease the dataset. The AQI graph was obtained and tested once we have reduced the dataset.

4.1 Model Validation

To validate and check the accuracy of the AQI prediction system using SVM, we have used two statistical means given below.

Calculated Mean Squared Error: Mean squared error (MSE) is a frequently used metric to assess how accurately a machine learning model performs when applied to regression issues, such as AQI prediction through SVM or SVR:

$$MSE = \frac{1}{n} * \sum_{k=0}^{n} (actual - predicted)^2 \tag{1}$$

where n is the total number of parameters, *actual* is the observed y-value, and *predicted* is the y-value from regression.

Calculated Root Mean Squared Error: The root mean squared error (RMSE) metric is a commonly used metric to evaluate the performance of a regression model:

$$RMSE = \sqrt{\frac{1}{n} * \sum_{k=0}^{n} (actual - predicted)^2} \tag{2}$$

where n is the total number of parameters, *actual* is the observed y-value, and *predicted* is the y-value from regression.

To keep the results organized and to make a graph, we constructed a box plotting graph. The graph shown in Fig. 6 indicates that the line has some degree of variation in it.

From this data study, we learned that hour and month formats are the most effective ways to resample this data. We obtained the reduced matrix for this purpose, which was then utilized to fit the data to our model.

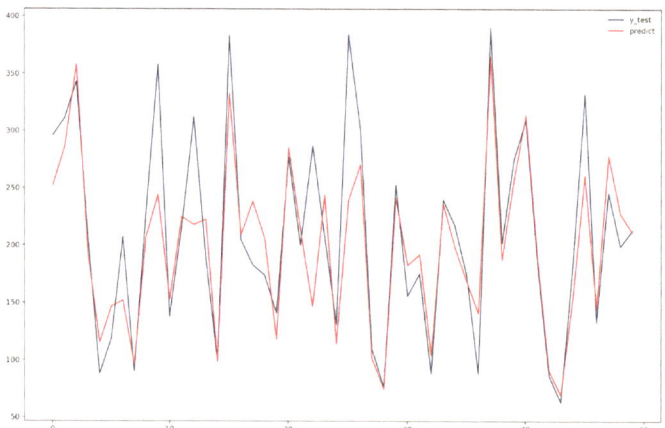

Fig. 6 Graph of the final prediction of AQI

5   Conclusion

The main purpose of this study is to forecast the AQI, based on its past readings. This proactive strategy offers a chance for early detection of worsening air quality, enabling the government to quickly take appropriate action. This work demonstrates the value of machine learning algorithms as tools for accurate monitoring and air pollution detection. We have developed predictive approaches for detecting the air quality index by examining previous data and utilizing cutting-edge algorithms. With early detection systems, the government can quickly put into action steps to protect public health and reduce environmental harm, such as pollution limits, traffic management, and public health advisories. The effective use of machine learning algorithms in this work indicates the growing importance of artificial intelligence as well as data-driven strategies to tackle challenging environmental concerns. Recent improvements in machine learning approaches have shown a promising boost in AQI prediction model accuracy.

References

1. K.B. Shaban, A. Kadri, E. Rezk, Urban air pollution monitoring system with forecasting models. IEEE Sens. J. vol. **16**(8), 2598–2606 (2016). https://doi.org/10.1109/JSEN.2016.2514378
2. S.K. Bamrah, K.R. Saiharshith, K.S. Gayathri, Application of random forests for air quality estimation in india by adopting terrain features, in *4th International Conference on Computer, Communication and Signal Processing (ICCCSP)* (Chennai, India, 2020), pp. 1–6. https://doi.org/10.1109/ICCCSP49186.2020.9315252
3. T.M. Amado, J.C.D. Cruz, Development of machine learning-based predictive models for air quality monitoring and characterization, in *TENCON IEEE Region 10 Conference* (Jeju, Korea (South), 2018), pp. 0668–0672. https://doi.org/10.1109/TENCON.2018.8650518
4. H. Srivastava, G.K. Sahoo, S.K. Das, P. Singh, Performance analysis of machine learning models for air pollution prediction, in *International Conference on Smart Generation Computing, Communication and Networking (SMART GENCON)* (Bangalore, India, 2022), pp. 16. https://doi.org/10.1109/SMARTGENCON56628.2022.10084037
5. K.M.O.V.K. Kekulanadara, B.T.G.S. Kumara, B. Kuhaneswaran, Machine learning approach for predicting air quality index, in *International Conference on Decision Aid Sciences and Application (DASA)* (Sakheer, Bahrain, 2021), pp. 622–626. https://doi.org/10.1109/DASA53625.2021.9682221
6. J.L. Qin et al., An improved novel nonlinear algorithm of area-wide near-surface air temperature retrieval. IEEE J. Sel. Top. Appl. Earth Obs. Remote. Sens. **11**(3), 830–844 (2018). https://doi.org/10.1109/JSTARS.2018.2793846
7. B. Liu, W. Yu, Y. Wang, Q. Lv, C. Li, Research on data correction method of micro air quality detector based on combination of partial least squares and random forest regression. IEEE Access **9**, 99143–99154 (2021). https://doi.org/10.1109/ACCESS.2021.3096216
8. C. Li, Y, Li, Y, Bao, Research on air quality prediction based on machine learning, in *2nd International Conference on Intelligent Computing and Human-Computer Interaction (ICHCI)* (Shenyang, China, 221), pp. 77–81. https://doi.org/10.1109/ICHCI54629.2021.00022
9. P. Bhalgat, S. Bhoite, S. Pitare, Air quality prediction using machine learning algorithms. Int. J. Comput. Appl. Technol. Res. **8**(9), 367–370 (2019)
10. A.G. Soundari, J.G. Jeslin, A.C. Akshaya, Indian air quality prediction and analysis using machine learning. Int. J. Appl. Eng. Res. **14**(11), 181–186 (2019)

11. U. Mahalingam, K. Elangovan, H. Dobhal, C. Valliappa, S. Shrestha, G. Kedam, A machine learning model for air quality prediction for smart cities, in *International Conference on Wireless Communications Signal Processing and Networking (WiSPNET)* (Chennai, India, 2019), pp. 452–457. https://doi.org/10.1109/WiSPNET45539.2019.9032734
12. M. Kulkarni, A. Raut, S. Chavan, N. Rajule, S. Pawar, Air quality monitoring and prediction using SVM, in *6th International Conference On Computing, Communication, Control And Automation (ICCUBEA)* (Pune, India, 2022), pp. 1–4. https://doi.org/10.1109/ICCUBEA54992.2022.10010942
13. K.C. Reddy, K.N. Reddy, K.B. Prasad, P.S. Rajendran, The prediction of quality of the air using supervised learning, in *6th International Conference on Communication and Electronics Systems (ICCES)*, (Coimbatre, India, 2021), pp. 1–5. https://doi.org/10.1109/ICCES51350.2021.9488983
14. L. Hota, P.K. Dash, K.S. Sahoo, A.H. Gandomi, Air quality index analysis of indian cities during covid-19 using machine learning models: A comparative study, in *8th International Conference on Soft Computing & Machine Intelligence (ISCMI)* (Cario, Egypt, 2021), pp. 27–31. https://doi.org/10.1109/ISCMI53840.2021.9654925

Fake News Detection Using Machine Learning Techniques

Lalita Kumari, Aakanksha Kumari, and Rahul Ahuja

Abstract The rise of social media and online news platforms has resulted in a concerning surge in the dissemination of false information. This trend poses a significant danger to open dialogue, democratic processes, and societal cohesion. Utilizing machine learning and deep learning methodologies has shown promise in the fight against fake news. This article presents a thorough examination of the current advancements in identifying fake news through deep learning and other machine learning techniques. Various methods of extracting features and classifying information in this field are discussed, along with an evaluation of different models' performance on established datasets. Furthermore, this paper identifies the obstacles and constraints of present strategies and delves into potential avenues for future research.

Keywords Fake news · Deep learning · Machine learning

1 Introduction

In the contemporary era of digital advancement, the swift distribution of information via social media platforms and online news outlets has posed an unparalleled challenge in the identification of credible and dependable sources. The rise of fabricated news, meticulously designed to mislead or influence public perceptions, represents a significant menace to rational decision-making and societal cohesion. To address this growing concern, this study investigates the application of machine learning and

L. Kumari (✉) · A. Kumari · R. Ahuja
Amity School of Engineering and Technology, Patna, India
e-mail: kumaril2003@yahoo.co.in

A. Kumari
e-mail: aakansha.7079@gmail.com

R. Ahuja
e-mail: rahuja@ptn.amity.edu

deep learning techniques for identifying fake news across social media platforms and online news articles. Our investigation employs the liar dataset, a comprehensive compilation of sentences classified as true or false, supplemented by speaker affiliations. Comprising 16 columns and 12788 rows, this dataset serves as a valuable resource for scrutinizing the linguistic and contextual indicators that differentiate spurious news from authentic content. Our process involves a holistic approach encompassing the stages of preprocessing, feature extraction, and model training. During the preprocessing phase, the focus is on cleansing and standardizing the text data. Furthermore, methods such as bag-of-words (BoW) and term frequency-inverse document frequency (TF-IDF) are employed to convert textual data into numerical representations, facilitating thorough analysis.

To effectively identify fake news, we train six different machine learning and deep learning models: Naive Bayes, logistic regression, random forest, decision tree, neural networks, and SVM. Naive Bayes serves as a baseline for evaluating the performance of other models, while neural networks, with their ability to capture complex patterns in data, are expected to achieve higher accuracy levels.

The efficacy of the neural network architecture is additionally enhanced by incorporating dimensionality reduction techniques, notably principal component analysis (PCA) and t-distributed stochastic neighbor embedding (t-SNE). These methods play a crucial role in diminishing the number of features, effectively eliminating extraneous information, and optimizing feature extraction. By preserving the overall data relationships, PCA and t-SNE allow for a more profound understanding and the attainment of highly precise outcomes. Our research significantly enhances the field of fake news detection by integrating a combination of machine learning and deep learning techniques, incorporating dimensionality reduction algorithms, and leveraging a diverse dataset. The outcomes of our investigation will guide the creation of more durable and efficient systems for spotting fake news, thereby advancing well-informed decision-making and the preservation of the credibility of online content.

2   Literature Review

The field of fake news detection has witnessed a surge in research endeavors, with scholars globally striving to combat the deleterious impact of misinformation on society. Models designed for fake news detection can assist journalists in directing their attention toward disseminating precise and dependable information. Machine learning (ML) and deep learning (DL) methodologies utilized in fake news detection have been reviewed extensively in [1, 2], which highlight the potential of DL approaches and propose an LSTM-based model with a 94% accuracy in detecting fake news.

The paper also underscores the potential expansion of the model to encompass other domains, including social media posts and online reviews.

The study presents an integrated approach that amalgamates linguistic attributes, user engagement behaviors, and network analytics to effectively discern and categorize deceptive or fabricated news articles. The primary discovery of this study indicates that the suggested method adeptly identifies false information, offering significant implications for crafting automated solutions and tactics to counteract misinformation across social media platforms. It offers a summary of the methodologies employed in spotting fake news, encompassing natural language processing, machine learning, and deep learning frameworks. The article underscores the hurdles encountered in fake news detection, including the evolution of adversarial methods and ethical dilemmas, stressing the necessity of comprehending these obstacles to devise viable remedies.

Text-based fake news can be effectively classified. This research specifically concentrates on categorizing fake news shared on social media platforms using conventional approaches and various classifiers, achieving an accuracy rate ranging from 81 to 100%. It is feasible to classify fake news shared on social media, particularly through the utilization of a convolutional neural network.

Machine learning is playing a vital role in the classification of information. The paper discusses the challenge of distinguishing between false and true information on social media, the research challenge of automatically checking information for categorizing it as false or true, and the role of machine learning in this classification, as well as the review of various ML approaches in the fake/fabricated news detection, and the potential improvisation through implementing deep learning.

A range of machine learning and deep learning techniques has been explored for fake news detection. Researchers conducted comprehensive studies, with Prachi achieving high accuracy using LSTM and BERT [3], and Alghamdi comparing the performance of various ML and DL techniques across different datasets [4]. Kumar (2021) [5] also reviewed these techniques, highlighting the challenges in distinguishing between real and fake news. Jaybhaye (2023) [1] emphasized the importance of these models in verifying news authenticity. These studies collectively underscore the potential of machine learning and deep learning [6] in addressing the pervasive issue of fake news.

3 Methodology

The proposed approach for identifying fake news is delineated in Fig. 1, which comprises multiple steps.

Dataset Used

We utilized the "Liar Dataset," which comprises sentences along with information about their speakers and affiliations, with labels indicating whether the statements are associated with fake news or legitimate content. The dataset consists of 16 columns and 12788 rows, providing a substantial volume of data for analysis.

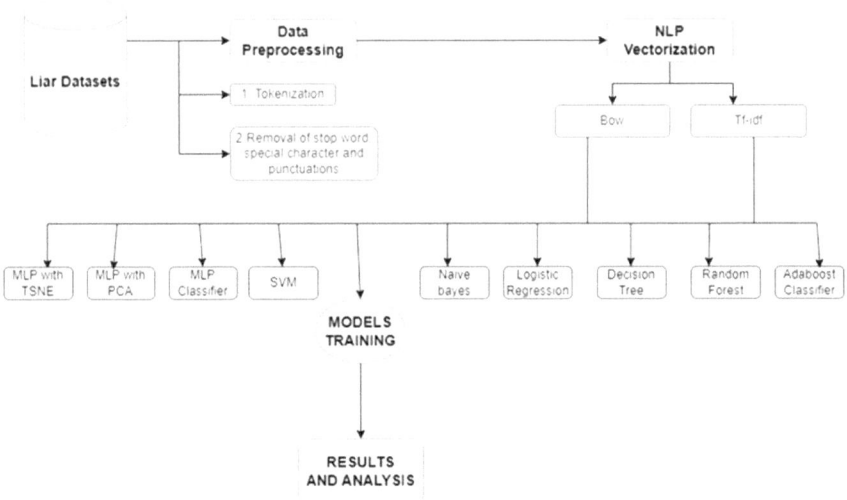

Fig. 1 Flowchart of the proposed model

Data Preprocessing

Data preprocessing played a vital role in preparing the dataset for analysis in our research. As shown in Fig. 2, preprocessing involved several key tasks:

- Eliminating punctuation marks, white spaces, and unnecessary characters from the text data.
- Employing NLP techniques to tokenize the text data, breaking sentences down into individual words or tokens.
- Utilizing term frequency-inverse document frequency (TF-IDF) and bag-of-words (BoW) methodologies to transform tokenized data into numerical representations.
- Generating word clouds to visualize word frequencies.
- Label encoding the "Political Party" and "Speaker" columns to convert categorical data into numerical form.

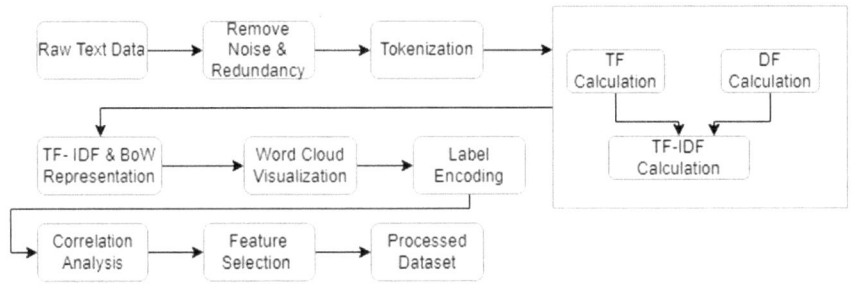

Fig. 2 Steps for data preprocessing

- Selectively removing 12 out of 16 columns based on a heatmap analysis that highlighted columns with strong correlations, which were subsequently excluded.

NLP Vectorization

We employed TF-IDF and BoW in natural language processing (NLP) vectorization, which helped convert the textual data into numerical features. TF represents the frequency of a term appearing within a document, whereas IDF gauges the importance of a term across the entire corpus. TF-IDF is computed as the product of TF and IDF.

t-SNE

To reduce the dimensionality of the dataset, we performed t-distributed stochastic neighbor embedding (t-SNE). This technique helps reduce the number of features, filtering out noise and facilitating better feature extraction. t-SNE preserves global data relationships, allowing for more accurate results:

$$C = \sum iKL(Pi\|Qi) \tag{1}$$

where KL represents the Kullback–Leibler divergence and P_i refers to the probability distribution over pairs of high-dimensional points. This distribution represents the likelihood of encountering specific pairs of points within the high-dimensional space. Q_i represents the probability distribution over pairs of low-dimensional points. This distribution signifies the likelihood of encountering particular pairs of points within the low-dimensional space. The algorithm progressively adjusts the positions of points in the lower dimensional space through iterative iterations of gradient descent, aiming to minimize the associated cost function.

Hyperparameter Tuning

We employed grid search to determine the best hyperparameters for our machine learning models. This process involved an exhaustive search over a range of hyperparameter values to identify the optimal configuration.

Machine Learning Models

We conducted experiments using a variety of machine learning models, such as support vector machines (SVM), neural networks, naive Bayes, logistic regression, decision trees, random forest, and AdaBoost classifier. For each model, we reported the accuracy scores and loss curves, providing insights into their performance.

Table 1 Accuracy scores of fake news detection by different ML algorithms

Model	TF-IDF without speaker and party	BoW without speaker and party	TF-IDF with speaker and party	BoW with speaker and party
Naive Bayes	58.10	54.59	60.1	54.39
Logistic regression	59.57	58.405	60.98	59.34
Random forest	59.60	60.61	61.05	62.89
Decision tree	56.89	57.22	58.94	59.89
SVM	59.35	58.41	58.31	58.10
AdaBoost classifier	57.41	58.17	60.42	60.75
MLP classifier	58.94	59.66	57.39	59.41
MLP with PCA	57.69	59.43	57.94	58.56
MLP with TSNE	55.28	56.92	54.74	59.35

4   Results and Discussion

In this investigation, we utilized an array of machine learning techniques along with various data setups, encompassing BoW and TF-IDF, while sometimes accounting for speaker and party associations [7]. Table 1 presents the summary of accuracy scores achieved for each model and setup.

5   Conclusion

In conclusion, our study highlights the complexity of fake news detection and the multifaceted impact of different features and vectorization techniques on model performance. While certain models, notably random forest, exhibited promising results, the variability in accuracy across different configurations emphasizes the need for tailored approaches for specific datasets and contexts.

Furthermore, the inclusion of speaker and party affiliations showed mixed results, suggesting that the influence of these features on fake news detection varies depending on the algorithm used. This research offers significant insights into both the challenges and opportunities within the realm of fake news detection, thereby laying a foundation for future investigations to delve into novel features, advanced algorithms, and innovative methodologies aimed at bolstering the precision and dependability of fake news detection systems.

References

1. S.M. Jaybhaye et al., Fake news detection using LSTM based deep learning approach, in *ITM Web of Conferences* (2023), n. pag. https://doi.org/10.1051/itmconf/20235603005
2. M. Shetty et al., Automated method for fake news detection using machine learning, in *2023 International Conference on Network, Multimedia and Information Technology (NMITCON)* (2023), pp. 1–6. https://doi.org/10.1109/NMITCON58196.2023.10275880
3. N.N. Prachi et al., Detection of fake news using machine learning and natural language processing algorithms. J. Adv. Inf. Technol. (2022). https://doi.org/10.12720/jait.13.6.652-661
4. J. Alghamdi et al., A comparative study of machine learning and deep learning techniques for fake news detection. Inf. **13**, 576 (2022). https://doi.org/10.3390/info13120576
5. S. Kumar, T. Jyoti, A review: machine learning approach and deep learning approach for fake news detection. Int. J. Emerg. Trends Eng. Res. (2021). https://doi.org/10.30534/ijeter/2021/01982021
6. L. Kumari, P.C.N. Sharma, Measuring the effectiveness of deep learning in STEM education: a comparative analysis of student outcomes and engagement, in *STEM: A Multi-Disciplinary Approach to Integrate Pedagogies, Inculcate Innovations and Connections* (2023), pp. 163–174. https://doi.org/10.52305/UXLT2425
7. P. Santhiya, S. Kavitha, T. Aravindh, S. Archana, A.V. Praveen, Fake news detection using machine learning, in *2023 International Conference on Computer Communication and Informatics (ICCCI)* (Coimbatore, India, 2023), pp. 1–8. https://doi.org/10.1109/ICCCI56745.2023.10128339

A Controlled Exploration of Vulnerability Assessment and Penetration Testing Through Burp Suite

Preetish Ranjan, Santosh Kumar, Nitish Ojha, Umesh Kumar, Ritesh Ravi, Naveen Kumar, and Fakhruddin Khan

Abstract This paper is trying to present flaws by penetrating into the web application. Burp Suite is a powerful tool which is being used to penetrate and display all the proprietary information of an organization. The aim of this paper is to access the security issues and create awareness among public as to how information could be leaked and misused by an attacker. There are few organizations who are operating at a larger level in a public domain, but they are unaware of the fact that their employees' personal information is also public. A website is the first face of any organization; once compromised, it ultimately affects the brand value which takes years to build upon.

Keywords VAPT · Burp suite

P. Ranjan (✉) · U. Kumar · R. Ravi · N. Kumar · F. Khan
Amity University, Patna, India
e-mail: pranjan@ptn.amity.edu

U. Kumar
e-mail: ukumar@ptn.amity.edu

R. Ravi
e-mail: rravi@ptn.amity.edu

N. Kumar
e-mail: nkumar3@ptn.amity.edu

F. Khan
e-mail: fkhan@ptn.amity.edu

S. Kumar
Nalanda College of Engineering, Chandi (Nalanda), Bihar, India
e-mail: santoshrathore.kumar20@gmail.com

N. Ojha
Amity University Noida, Noida, India
e-mail: nkumar30@amity.edu

1   Introduction

There are many websites which are being operated without proper security assessment. The users believe that who will attack them and what benefit any attacker will have by attacking their website. Without proper vulnerability assessment and penetration testing, a website is launched. But they do not know that attackers may not attack them directly, but they can attack indirectly. Most of the time, every cyber-attack has some hidden financial motive. As per Techpedia, 493.33 million ransomware attacks were detected worldwide, and the global average data breach cost was $4.35 million in the year 2022. In these kinds of attacks, attackers access the database, encrypt the data, and then ask for ransom to provide the key. The situation becomes worse if the website or web application belongs to hospitals or some health agency. There are many other organizations such as banks and railways which provide critical service to society where data plays a very important role. The proxy tool in the Burp Suite is a crucial component for intercepting and analyzing HTTP and HTTPS traffic between web browser and the target application by enabling a tab Intercept. While intercepting HTTPS traffic, Burp Suit provides a built-in certificate. Requests can be modified before they are sent to the server. They can even be forwarded to the server by clicking the "Forward" button without the modification of the request. This feature can be used in a positive sense too for testing and accessing the security of the application. There are "Repeater" and "Intruder" tools for further detailed analysis. Therefore, Burp Suite is an essential component for security testing for identifying vulnerabilities and security issues in a web application. It is always important to have permission to test the target application as unauthorized access can be illegal and unethical.

2   Literature Review

Sangeeta et al. explored Burp Suite as an automatic software testing tool for a web application. It works as a MITN attack vector browser and target web application's request and responses to deduce the vulnerabilities. Burp Suite creates several requests and analyzes the response to collect and produce the proof of vulnerabilities. They also worked on various other tools which are manual, automated, open source, or commercial with different functionalities and applicability. They also addressed the strategies to choose one of the best VAPT tools. They concluded that none of the tools is self-sufficient to identify the security risk in a web application. However, this paper provides comparative and collective analysis of web application VAPT [1]. Keyur et al. first explored OWASP top 10 vulnerabilities and then various tools available for VAPT. They observed that SQL injection is the most frequent attack in comparison to broken authentication and session management. While exploring the automated VAPT tools such as NMAP, Nessus, Burp Suite, Accuentix, Metaspoilt, Harvester, Wireshark, ZAZ, Beef, and SQLMAP, they also observed the disadvantage of these tools that they are unable to identify the logical attacks and only generate the analysis according to the policies defined by the tester and might give false positive results [2]. Riccardo et al. performed an exploratory controlled experiment, in which nine participants analyzes the security of two open-source blogging applications. They concluded that static analysis found more vulnerabilities than penetration testing. They focused on the poor programming practices that enabled potential attacks. Most frequent poor practices involve loops that are incorrectly iterated over the elements of an array, opening the possibility of a buffer overflow. They implemented Fortify's Static Code analyzer and Burp Suite as the tool to support the penetration testing process. However, static and dynamic tools analysis found different vulnerabilities in the same part of code [3]. Cyber-space users are spending much of their time either with some electronic gadgets or with web applications or web portals or social media platforms. Therefore, they are more prone to be attacked. Information is viral due to some genuine reason, but sometimes they are made viral with false content [4]. This can result in a change in the sentiment and can affect the decision-making ability of the user. Cyber-attacks are extending their reach to the mass level and affecting society. Eventually, this can radicalize the society toward particular issue and can result in the devastating impact in society [5].

3   Methodology

The very first step is the identification of target web applications. The proxy service of Burp Suite acts as a gateway between a client and another server. Proxies serve various purposes, including enhancing security, improving the performance, and enabling anonymity. HTTP/HTTPS proxies can filter and inspect incoming and outgoing traffic, blocking malicious content, viruses, and malware. Organizations and educational institutions often use SOCKS proxies to restrict access to specific websites or content categories. This helps to enforce the Internet usage policies and prevent employees or students from accessing inappropriate or non-work-related material. Proxies can distribute incoming requests to multiple servers, ensuring a balanced workload and improved performance. This is crucial for high-traffic websites and online services. A proxy can mask the real IP address, making it appear as if requests are coming from a different location or device. This can help to protect privacy and identity. Proxies can also store copies of frequently accessed web pages, images, or other content. When a user requests the same content, the proxy can deliver it from its cache, reducing the load on the source server and speeding up the content delivery.

If the proxy is not set by default, then it must be set to 127.0.0.1 through advanced network setting, and the default port for Burp Suite is 8080. The choice of a proxy service and its configuration depends on specific needs and use case, whether it is for personal privacy, security, accessing geo-restricted content, or improving network performance.

The second step is to enable the intercept by Intercept which is on a tab as shown in Fig. 1 which displays the traffic. The modification of web traffic is a valuable tool

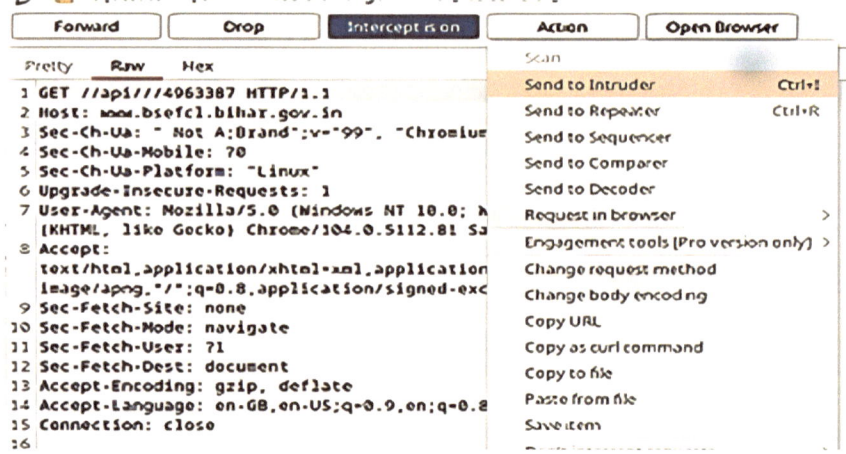

Fig. 1 Interception of traffic

for security testing and analysis; it should only be performed on web applications and websites with explicit permission to test. Unauthorized interception and modification of web traffic is illegal and unethical.

The third step is that the entire information is shared to the intruder as shown in Fig. 1.

This tool allows to automate and perform various attacks on web applications. The request to be tested is located and sent to the intruder. There are different sections for configuration:

Positions: Position is specified to define the parts of the request to attack using placeholders (§1§, §2§, etc.).

Payloads: It is the data to be used in the attack which can be predefined or customized.

Options: The available attack type (e.g., Sniper, Battering Ram, Pitchfork), number of threads, and more.

Grep-Match: It defines patterns to look for in the responses to identify successful attacks.

Successful attacks and potential vulnerabilities will be highlighted as displayed in Fig. 2, and it can be exported for further analysis. Proper authorization before using the intruder tool to test web applications is always required; otherwise, it can have legal consequences.

Burp Suite's scanner tool can automatically scan a web application for a wide range of vulnerabilities, including SQL injection, cross-site scripting (XSS), and more. Weak or default credentials can be tested by performing a brute force attack on login forms. The parameters present in different payload positions can be altered within requests to test for issues like parameter tampering or business logic vulnerabilities. These kinds of attacks are known as cluster bomb attack. Multiple attacks using different payloads and positions within a single request are known as pitchfork

Fig. 2 Sniper attack started on a website

attacks. It also facilitates battering RAM attacks for single payload against multiple positions within requests. These attacks can be launched manually to identify potential vulnerabilities, perform parameter manipulation, or test different scenarios. The quality of randomness in tokens, session IDs, or other values generated by the web application can also be analyzed. Data can also be decoded and encoded in various formats, such as Base64, URL encoding, and more.

4 Results and Discussion

Proxies record and monitor network traffic, providing valuable insights into the activities of users, potential security threats, and network performance. After starting the proxy listener, the website or web application is visited which is to be analyzed. This will pause the traffic between the browser and the web server. These requests can either be modified before they are sent to the server or forwarded without changes. These responses can contain valuable information for security testing and analysis. Figure 3 displays the information gathered as the tab Intercept which is on is clicked to intercept the traffic of the web application.

General information about the web application:

GET is the method for HTTP request from the resource/api/4963367 where HTTP/ 1.1 is the version of HTTP protocol used by the host.

Host Information:

A web application hosted in remote server can detect user's CPU architecture and operating system by using user-agent client hints (UA-CH). The user-agent client hints format is used by browsers to provide user agent information to websites. It

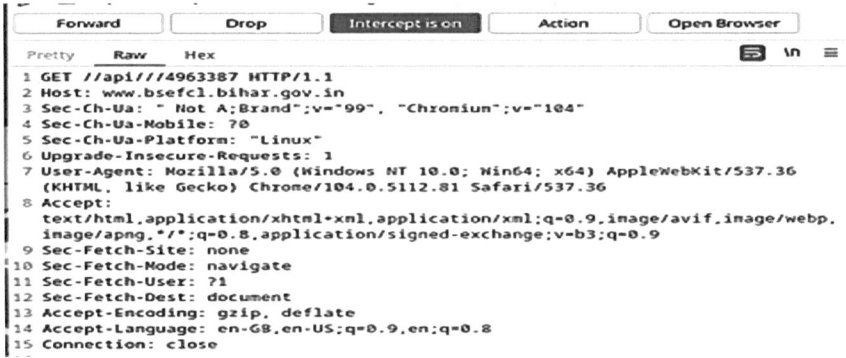

```
   Forward          Drop       Intercept is on        Action       Open Browser

Pretty    Raw    Hex                                              ⊟  \n  ≡
 1 GET //api///4963387 HTTP/1.1
 2 Host: www.bsefcl.bihar.gov.in
 3 Sec-Ch-Ua: " Not A;Brand";v="99", "Chromium";v="104"
 4 Sec-Ch-Ua-Mobile: ?0
 5 Sec-Ch-Ua-Platform: "Linux"
 6 Upgrade-Insecure-Requests: 1
 7 User-Agent: Mozilla/5.0 (Windows NT 10.0; Win64; x64) AppleWebKit/537.36
   (KHTML, like Gecko) Chrome/104.0.5112.81 Safari/537.36
 8 Accept:
   text/html,application/xhtml+xml,application/xml;q=0.9,image/avif,image/webp,
   image/apng,*/*;q=0.8,application/signed-exchange;v=b3;q=0.9
 9 Sec-Fetch-Site: none
10 Sec-Fetch-Mode: navigate
11 Sec-Fetch-User: ?1
12 Sec-Fetch-Dest: document
13 Accept-Encoding: gzip, deflate
14 Accept-Language: en-GB,en-US;q=0.9,en;q=0.8
15 Connection: close
```

Fig. 3 Information gathered

also reveals the browser brand, version number, and the device platform on which the browser is running.

Sec-Ch-Ua-Mobile: ?0 header indicates that the application has not been accessed through the mobile device.

Sec-Ch-Ua-Platform: "Linux" header indicates that Linux is the operating system which is used by the user for access.

Upgrade-Insecure-Requests: 1 is the attribute of the meta tag which directs the domain to migrate from HTTP to HTTPS especially non-navigational links of images present in the websites.

This web application is compatible with several other browsers such as Mozilla/5.0, ApplewebKit, Chrome, and Safari.

Accept reveals the list of header requests which are accepted by the server hosting that web application.

Sec-Fetch-Site: none tells that the request is user originated either by entering a URL into the address bar or by opening a bookmark, or by dragging-and-dropping into the browser window.

Sec-Fetch-Mode: navigate implies that the request is initiated through the navigation among HTML documents.

Sec-Fetch-User: ?1 is the header used by the server to ensure that the request is originated by the user and not by any other document or iframe, etc.

Sec-Fetch-Dest: document header directs the server on how the expected response for the particular service request is to be used. If the request is for audio or video destination, then the response will be in a corresponding format.

Accept-Encoding: gzip, deflate list the media-types that the browser can accept.

Accept-Language indicates the language which a client aspect to receive the response.

Connection: close indicates that either the client or the server wants to close the connection.

Responses can be compared to identify differences, which can be useful for detecting variations in error messages or other sensitive information. The functionality of Burp Suite can be extended to create an attack or automate specific tasks. The Spider tool is available to automatically crawl the application, discover content, and create a site map. Burp Suite's content discovery tools can be used to find hidden files and directories on the web server. It can also scan the target server for open ports and services to identify potential attack surfaces. These are some of the primary attacks and testing techniques available in Burp Suite.

Snipper type of attack has been launched in a government website. Snipper attack uses a set of payloads which have the wordlist which attacks the common vulnerabilities like "SQL injection and XSS". The word from the wordlist is passed as a payload username = $username$ and password = $password$. Sniper brute forces each position with the payload turn by turn as shown in Fig. 4.

All the personal information of the employee is clearly visible in Fig. 5. It can be misused by the attacker in a different capacity. If the employee's personal information such as Aadhar number, address, and PAN information have been accessed by attackers, they can misuse this to reach them through phishing, spamming, mail, or SMS.

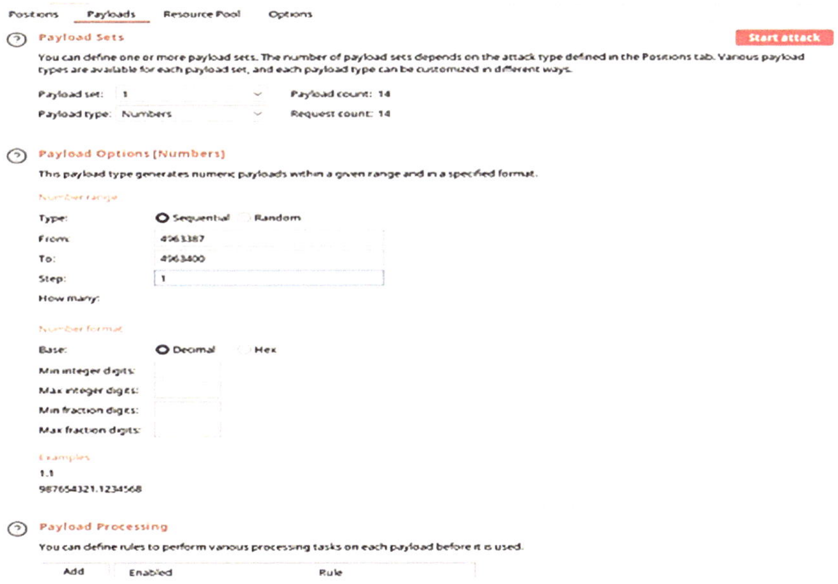

Fig. 4 Setting of payload

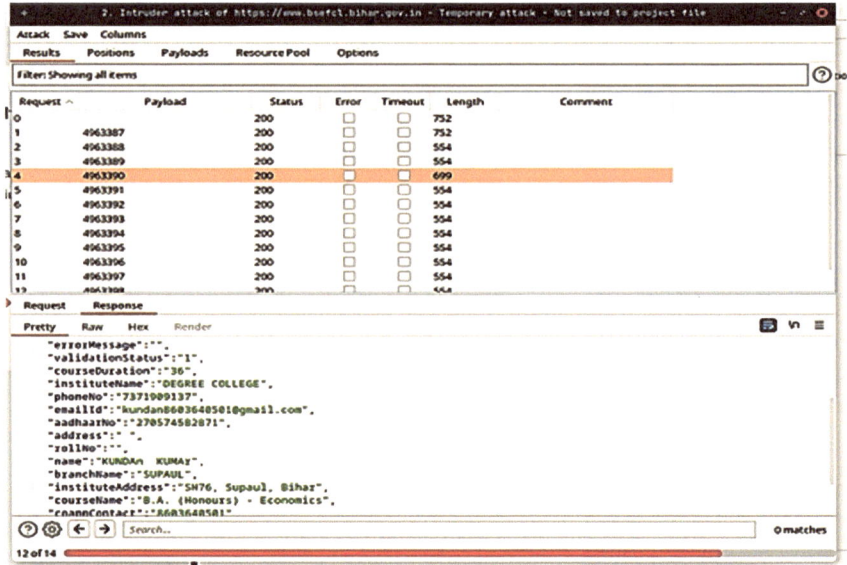

Fig. 5 Personal information of an employee

5 Conclusion

Burp Suite is a group of tools which includes proxy server, intruder, spider, etc. All the tools are basically based on the man-in-the-middle-attack vector which tries to interrupt the traffic between the browser and target application under a scanner through Burp Suite. It is also an automatic and passive scanning tool which analyzes the contents of requests of the client made to the server and responses received by the client from the server. After analyzing these requests and the responses received, Burp Suite creates its own requests and analyzes the responses to access the vulnerabilities. This information can be used for phishing or stalking which can cause financial loss or can be sometimes a threat to life.

References

1. S. Nagpure, S. Kurkure, Vulnerability assessment and penetration testing of web application, in *3rd International Conference on Computing and Automation (ICCUBEA)* (IEEE Xplore, 2017)
2. K. Patel, A survey on vulnerability assessment and penetration testing for secure communication, in *Proceedings of the Third International Conference on Trends in Electronics and Informatics (ICOEI)* (IEEE Xplore, 2019)
3. R. Scandatiato, J. Walden, W. Joosen, Static analysis versus penetration testing: a controlled experiment, in *IEEE 24th International Symposium on Software Reliability Engineering (ISSRE)* (2013)

4. P. Ranjan, A. Vaish, Socio-technical attack approximation based on structural virality of information in social networks, in *International Journal of Information Security and Privacy (IJDCF)* (IGI Global, 2021)
5. P. Ranjan, V. Singh, P. Kumar, S. Prakash, Models for the detection of malicious intent people in society, Int. J. Digit. Crime Forensics (2018). IGI Global